Synthesis Lectures on Engineering, Science, and Technology

The focus of this series is general topics, and applications about, and for, engineers and scientists on a wide array of applications, methods and advances. Most titles cover subjects such as professional development, education, and study skills, as well as basic introductory undergraduate material and other topics appropriate for a broader and less technical audience.

Mário Saldanha · Gustavo Sanchez ·
César Marcon · Luciano Agostini

Versatile Video Coding (VVC)

Machine Learning and Heuristics

 Springer

Mário Saldanha
Instituto Federal Sul-rio-grandense
Pelotas, Rio Grande do Sul, Brazil

César Marcon
Pontifícia Universidade Católica do Rio
Grande Sul
Porto Alegre, Rio Grande do Sul, Brazil

Gustavo Sanchez
Instituto Federal do Sertão de Pernambuco
Salgueiro, Pernambuco, Brazil

Luciano Agostini
Universidade Federal de Pelotas
Pelotas, Rio Grande do Sul, Brazil

ISSN 2690-0300   ISSN 2690-0327   (electronic)
Synthesis Lectures on Engineering, Science, and Technology
ISBN 978-3-031-11642-1   ISBN 978-3-031-11640-7   (eBook)
https://doi.org/10.1007/978-3-031-11640-7

This Springer imprint is published by the registered company Springer Nature Switzerland AG
The registered company address is: Gewerbestrasse 11, 6330 Cham, Switzerland

Foreword

Versatile Video Coding (VVC), also known as ITU-T Recommendation H.266 | ISO/IEC 23090-3, is the latest international video coding standard developed jointly by the ITU-T Video Coding Experts Group (VCEG) and ISO/IEC Moving Picture Experts Group (MPEG). Finalized in July 2020, VVC represents the fruitful outcome of years of research and development, offering much improved compression efficiency than its predecessor H.265/HEVC and greater versatility to support a broad range of applications, e.g., high-definition video and beyond, screen content video, high dynamic range video, and 360 immersive video. The high compression efficiency of VVC is enabled by a large number of advanced intra- and inter-picture prediction techniques, which pose a great challenge on its encoder implementation, software, and hardware, when it comes to choose the best combination of coding options with limited resources.

This book addresses expressly efficient encoding solutions for intra-picture prediction in VVC. It features an in-depth characterization of intra-picture prediction tools in terms of their impact on compression performance and the encoding runtime. An insightful summary of the findings motivates a series of novel ideas on fast yet efficient encoding solutions. These begin with a conventional, statistical approach to fast multi-type tree partitioning decision for luminance blocks, followed by a more evolved learning-based method that trains Light Gradient Boosting Machine classifiers to decide on the tree-structure split type. The high correlation of the split types between the luminance and chrominance components is further exploited in another heuristic attempt to reduce encoding effort. The book also covers fast intra-prediction mode decision and fast transform-type selection, proposing methods that operate decision tree classifiers in a cascading manner to avoid unnecessary coding option evaluation. Special emphasis is put on the effectiveness of these simple classifiers in striking a good balance between complexity and accuracy in the context of fast mode decision. The hand-crafted delicate features present another highlight of the book.

The authors of the book are experienced experts in this area who have made significant contributions to developing fast mode decision algorithms for various ITU-T and ISO/IEC video coding standards. Methods presented in the book have been scientifically validated in VVC Test Model (VTM). Overall the book is a good read. Some unique views are nowhere to be found. It is highly recommended to students, researchers, and engineers who like to grasp quickly various opportunities to develop fast VVC encoding algorithms beyond existing methods in the literature.

Wen-Hsiao Peng
National Yang Ming Chiao Tung University
Taipei, Taiwan

Acknowledgements

The authors thank their institutions in Brazil for supporting the development of this work, including the Federal University of Pelotas (UFPel), the Federal Institute of Education Science and Technology Sul-Rio-Grandense (IFSul), the Federal Institute of Education, Science and Technology of Sertão Pernambucano (IFSertãoPE), and the Pontifical Catholic University of Rio Grande do Sul (PUCRS). The authors also thank the Brazilian research support agencies that financed this research: the Foundation for Research Support of Rio Grande do Sul (FAPERGS), the National Council for Scientific and Technological Development (CNPq), and the Coordination for the Improvement of Higher Education Personnel (CAPES).

Contents

1 Introduction 1
    References 4

2 Versatile Video Coding (VVC) 7
    2.1   Basic Video Coding Concepts 7
    2.2   VVC: A Hybrid Video Encoder 8
    2.3   VVC Frames Organization and Block Partitioning 10
    2.4   VVC Encoding Tools 13
           2.4.1   VVC Prediction Tools 14
           2.4.2   VVC Residual Coding and Entropy Coding 16
           2.4.3   VVC In-Loop Filters 17
    2.5   VVC Common Test Conditions 18
    References 20

3 VVC Intra-frame Prediction 23
    3.1   Angular Intra-prediction 25
    3.2   Multiple Reference Line Prediction 26
    3.3   Matrix-Based Intra-prediction 27
    3.4   Intra-sub-partition 28
    3.5   Encoding of Chrominance CBs 29
    3.6   Transform Coding 30
    References 32

4 State-of-the-Art Overview 35
    References 40

5 Performance Analysis of VVC Intra-frame Prediction 43
    5.1   Methodology 43
    5.2   VVC Versus HEVC: Intra-frame Compression Performance
           and Computational Effort Evaluation 44
    5.3   VVC Intra-frame Computational Effort Distribution of Luminance
           and Chrominance Channels 45

5.4   VVC Intra-frame Block Size Analysis 46
5.5   VVC Intra-frame Encoding Mode Analysis 48
5.6   VVC Intra-frame Encoding Transform Analysis 51
5.7   Rate-Distortion and Computational Effort of VVC Intra-frame
       Coding Tools 56
5.8   General Discussion 60
References 61

6   **Heuristic-Based Fast Multi-type Tree Decision Scheme for Luminance** ... 63
6.1   Initial Analysis 63
6.2   Designed Scheme 65
6.3   Results and Discussion 66
References 69

7   **Light Gradient Boosting Machine Configurable Fast Block
    Partitioning for Luminance** 71
7.1   Background on LGBM Classifiers 71
7.2   Methodology 72
7.3   Features Analysis and Selection 74
7.4   Classifiers Training and Performance 75
7.5   Classifiers Integration 80
7.6   Results and Discussion 82
References 87

8   **Learning-Based Fast Decision for Intra-frame Prediction Mode
    Selection for Luminance** 89
8.1   Fast Planar/DC Decision Based on Decision Tree Classifier 90
8.2   Fast MIP Decision based on Decision Tree Classifier 91
8.3   Fast ISP Decision Based on the Block Variance 92
8.4   Learning-Based Fast Decision Design 94
8.5   Results and Discussion 94
References 96

9   **Fast Intra-frame Prediction Transform for Luminance Using Decision
    Trees** 99
9.1   Fast MTS Decision Based on Decision Tree Classifier 100
9.2   Fast LFNST Decision Based on Decision Tree Classifier 101
9.3   Fast Transform Decision Design 101
9.4   Results and Discussion 103
References 105

10 Heuristic-Based Fast Block Partitioning Scheme for Chrominance 107

    10.1 Chrominance CB Splitting Early Termination Based
          on Luminance QTMT 108

    10.2 Fast Chrominance Split Decision Based on Variance of Sub-blocks ... 109

    10.3 Fast Block Partitioning Scheme for Chrominance Coding Design 112

    10.4 Results and Discussion 115

    References 118

11 Conclusions and Open Research Possibilities 119

Index 123

Introduction

The experience of video coding immersive technologies such as 3D videos, 360° video, Virtual Reality (VR), Augmented Reality (AR), and Mixed Reality (MR) lead to continuous growth in the development of new devices capable of handling digital videos [1]. Both industry and academia are investing several efforts to handle different kinds of video content efficiently. Nowadays, end users can buy various video devices, including TVs, computers, notebooks, smartphones, video games, tablets, 360° cameras, and VR glasses. At the same time, there is a notable popularization of digital services that use video technologies, including Netflix, Amazon Prime Video, YouTube, Facebook, Instagram, and WhatsApp. This video demand increase is creating enormous pressure on the network infrastructure [2]. The COVID-19 pandemic has made this process even more notable, where streaming providers had to reduce the transmitted video quality to allow the quality of service [3].

The development of new advanced video coding techniques is essential to support this increasing demand for transmission and storage of videos, especially in a scenario where video resolutions and frame rates tend to continue growing and new immersive visual information technologies become increasingly available. The novel video coding techniques must consider several important aspects, such as reducing the size of the encoded video (bitstream), providing high video quality, achieving real-time processing, having low energy consumption, and using computational resources according to their availability. Therefore, industry and academia have carried out a high effort in providing new video coding technologies, generating a significant number of encoders, such as Advanced Video Coding (AVC) [4], High-Efficiency Video Coding (HEVC) [5], VP9 [6], Audio Video Coding Standard 2 (AVS2) [7], Audio Video Coding Standard 3 (AVS3) [8], AOMedia Video 1 (AV1) [9], and Versatile Video Coding (VVC) [10].

A group of very important encoders were standardized together by the International Organization for Standardization (ISO) and by the International Telecommunication

Union (ITU). The first commercial success of these encoders was the worldwide known MPEG-2 [11], which was launched in 1996. MPEG is the acronym for the ISO group of experts which contributed with the standard development, the Moving Picture Expert Group. This standard is the same named as H.262 by the ITU organization. The next success of the collaborative effort between ISO and ITU experts was the AVC [4], launched in 2003. ISO called this standard as MPEG-4 Part 10 and ITU called this standard as H.264. The AVC successor was the HEVC [5], which was launched in 2013 and named as H.265 by ITU and as MPEG-H Part 2 by ISO. The current ISO and ITU state-of-the-art video coding standard is the VVC [10], which was launched in 2020. This standard was named as H.266 by ITU and named as MPEG-I Part 3 by ISO.

MPEG-2 was very important to popularize digital videos. This standard was applied in many applications, but maybe the most famous were the Digital Video Disk (DVD) and the first standards of digital television broadcast, like the ones defined by the Digital Video Broadcasting (DVB) and the Advanced Television Systems Committee (ATSC). The significant improvement in the image quality, when compared with analog video format, spread the use of MPEG-2 and boost the development of other video coding standards.

The AVC is another example of commercial success of ISO and ITU video coding standards. Several consolidated platforms have widely adopted AVC, such as Netflix and YouTube. However, the requirement for Full-High Definition (FHD) and Ultra-High Definition (UHD) video content, immersive technologies, and advanced computer architectures for parallel processing demonstrated that AVC was not efficient enough in this new scenario. To handle this challenge, ISO and ITU experts developed the HEVC to support the growing demand for higher resolutions and immersive technologies. HEVC doubled the compression rate for the same video quality when compared to the AVC [5]. However, this improvement in the coding efficiency comes at the cost of encoding complexity increases of 3.2 and 1.2 times compared to H.264/AVC in All-Intra (AI) and Random-Access (RA) configurations, respectively [12]. Besides, HEVC was designed targeting the technological advances in the industry like the multicore processors and Graphics Processing Unit (GPU), introducing Tiles and Wavefront Parallel Processing (WPP) [5] to allow efficient parallelism in the encoding/decoding processes. With the growing demand for immersive technologies that provide experiences beyond 2D videos, experts have also developed HEVC extensions, such as MV-HEVC and 3D-HEVC [13].

The next video coder developed by ISO and ITU was the VVC. ISO Moving Picture Experts Group (MPEG) and ITU-T Video Coding Experts Group (VCEG) created the Joint Video Experts Team (JVET) to develop the VVC standard, which was established as a Final Draft International Standard (FDIS) in July 2020. JVET experts have designed VVC with a focus on significantly obtaining higher compression efficiency than HEVC and having high versatility for efficient use in a broad range of applications and different types of video content, including UHD, High-Dynamic Range (HDR), screen content, 360° video, and resolution adaptivity. Moreover, since the industry poorly adopted

HEVC because of the lack of a reliable and consolidated licensing structure, a new body called Media Coding Industry Forum (MC-IF) [14] was established to define a clear and reasonable licensing model for VVC.

VVC introduces several novel techniques and enhancements in all video coding flow, including block partitioning, intra- and inter-frame predictions, transform, quantization, entropy coding, and in-loop filters, intending to improve the coding efficiency. Chapter 2 of this book presents a theoretical background about VVC, discussing some basic aspects of video coding and the main novelties brought by this standard. At the end of this chapter, we describe the Common Test Conditions (CTC), which explain the methodology required to fairly evaluate new algorithms proposed for VVC. Chapter 3 details the VVC intra-frame prediction since it is the focus of this book. This chapter details the VVC intra-frame prediction encoding flow and the novel tools used at the prediction and transform encoding steps.

The insertion of these advanced tools improved the VVC encoding efficiency but raised the encoder complexity expressively compared to HEVC. Our analysis presented in [15] and the work of Bossen et al. [16] showed that the VVC Test Model (VTM) [17] demands about 27 and 8 times more computational effort than the HEVC Test Model (HM) [18] for AI and RA encoder configurations, respectively. VTM is a reference implementation of all encoding features defined for the VVC standard. It implements all coding tools defined for VVC along with some heuristics to reduce the encoder complexity.

VVC is a video coding standard that tends to be widely adopted by the industry due to its high encoding efficiency; however, the industry will still face the challenge of being capable of designing real-time VVC encoders and decoders. Therefore, efficient simplifications are needed to enable real-time processing with low energy consumption, but with minor losses in coding efficiency.

Several works in the literature propose solutions to reduce the HEVC encoding complexity. Some of these works present solutions to accelerate the block partitioning decision [19, 21]; in contrast, the works [22, 23] introduce algorithms to reduce the processing time of some specific intra- and/or inter-frame prediction mode decisions. Furthermore, other works focus on the complexity control for the HEVC encoding process [24, 25].

Although these solutions present outstanding results in the HEVC, they cannot be used directly in VVC since it introduces several new advanced tools and modifications in the coding structure, significantly changing the encoding time distribution and context compared to HEVC. A dense set of these changes have been made to intra-frame prediction, which is the focus of this book, where the computational effort increases by 27 times over HEVC [15].

A discussion of the most relevant published works in this subject is presented in Chap. 4 of this book, where a dense set of related works targeting the computational effort reduction of VVC encoders are presented and discussed.

Chapter 5 of this book presents an extensive performance analysis of VVC intra-frame prediction, showing encoding time and usage distribution analysis. At the end of this chapter, we bring some insights into the main possibilities for reducing VVC encoder complexity, and some of these insights are explored in the solutions presented in the next chapters.

Chapters 6–11 present heuristic and machine learning solutions we developed for providing encoding time reduction in VVC intra-frame prediction with minimal impact on coding efficiency.

Chapter 6 presents a fast decision scheme based on a heuristic for the block partitioning of luminance blocks. This scheme explores the selected intra-frame prediction modes and the variance of block samples to decide when to avoid some block size evaluations. This solution reduced the encoding time by 43.23% at the cost of a 0.80% Bjontegaard Delta Bit Rate (BD-BR) [26] increase, which is a metric used to evaluate the coding efficiency losses.

A configurable fast block partitioning solution based on machine learning for luminance blocks is presented in Chap. 7. In this case, a machine learning classifier called Light Gradient Boosting Machine (LGBM) was used to reduce the number of evaluated partitions. A set of LGBM classifiers was trained and validated using real digital video sequences, this solution reduced the encoding time from 35.22 to 61.34% with BD-BR increase from 0.46 to 2.43%.

Chapter 8 presents our machine learning approach to generate a fast decision scheme for intra-mode selection of luminance blocks. This scheme was modeled with two solutions using decision trees and one solution based on a heuristic. This scheme provided an encoding time reduction of 18.32% with a 0.60% BD-BR increase.

A machine learning solution was also developed to reduce the encoding time for intra-transform mode selection of luminance blocks and this solution is presented in Chap. 9. This solution was also modeled using two decision tree classifiers, reaching an encoding time reduction of 11% with an increase of 0.43% in BD-BR.

The last solution presented in this book is a heuristic-based fast block partitioning scheme for chrominance blocks. This solution explores the correlation between chrominance and luminance coding structures and the statistical information in the chrominance block samples. This solution is presented in Chap. 10, and it reduced the chrominance encoding time in 60.03% with a BD-BR impact of 0.66%.

Finally, Chap. 11 concludes this book and brings some insights for future research works on VVC encoding.

References

1. Sherman, W., & Craig, A. (2018). *Understanding virtual reality: Interface, application, and design* (2nd ed.). Morgan Kaufmann Publishers.
2. CISCO. Cisco Annual Internet Report (2018–2023) White Paper. 2020. Retrieved October, 2021, from https://www.cisco.com/c/en/us/solutions/collateral/executive-perspectives/annual-internet-report/white-paper-c11-741490.html.

3. Dhapola, S. (2020). COVID-19 impact: Streaming services to dial down quality as internet speeds fall. Indian Express, 2020. Retrieved October, 2021, from 2021.https://indianexpress. com/article/technology/tech-news-technology/coronavirus-internet-speeds-slow-netflix-hot star-amazon-prime-youtube-reduce-streaming-quality-6331237/.

4. Marpe, D., Wiegand, T., Sullivan, G. T., & H. (2006). 264/MPEG4 advanced video coding standard and its applications. *IEEE Communications Magazine, 44*(8), 134–143.

5. Sullivan, G. et al. (2012). Overview of the high efficiency video coding (HEVC) standard. *IEEE Transactions on Circuits and Systems for Video Technology (TCSVT), 22*(12), 1649–1668.

6. Mukherjee, D. et al. (2013). The latest open-source video codec VP9-an overview and preliminary results. In *Picture coding symposium (PCS)* (pp. 390–393).

7. He, Z. et al. (2013). Framework of AVS2-video coding. In *IEEE International Conference on Image Processing (ICIP)* (pp. 1515–1519).

8. Zhang, J. et al. (2019). Recent development of AVS video coding standard: AVS3. In *Picture coding symposium (PCS)* (pp. 1–5).

9. Chen, Y. et al. (2018). An overview of core coding tools in the AV1 video codec. In *Picture coding symposium (PCS)* (pp. 41–45).

10. Bross, B. et al. (2021). Overview of the versatile video coding (VVC) standard and its applications. *IEEE Transactions on Circuits and Systems for Video Technology (TCSVT), 31*(10), 3736–3764.

11. Tudor, P. N. (1995). MPEG-2 video compression. *Electronics and Communication Engineering Journal, 7*(6), 257–264.

12. Vanne, J. et al. (2012). Comparative rate-distortion-complexity analysis of HEVC and AVC video codecs. *IEEE Transactions on Circuits and Systems for Video Technology (TCSVT), 22*(12), 1885–1898.

13. Sullivan, G. et al. (2013). Standardized extensions of high efficiency video coding (HEVC). *IEEE Journal of Selected Topics in Signal Processing (J-STSP), 7*(6), 1001–1016.

14. Samuelsson, J. (2020). Media coding industry forum progress report. *SMPTE Motion Imaging Journal, 129*(8), 100–103.

15. Saldanha, M. et al. (2020). Complexity analysis of VVC intra coding. In *IEEE International Conference on Image Processing (ICIP)* (pp. 3119–3123).

16. Bossen, F. et al. (2021). AHG report: Test model software development (AHG3). In *JVET 23rd Meeting, JVET-W0003, Jul. 2021.*

17. Chen, J., Ye, Y., & Kim, S. (2020). Algorithm description for versatile video coding and test model 10 (VTM 10). In *JVET 19th Meeting, JVET-S2002, Teleconference, Jul. 2020.*

18. Rosewarne, C. et al. (2015). High efficiency video coding (HEVC) test model 16 (HM 16). Document: JCTVC-V1002, Geneva, CH, Oct. 2015.

19. Xu, M. et al. (2018). Reducing complexity of HEVC: A deep learning approach. *IEEE Transactions on Image Processing (TIP), 27*(10), 5044–5059.

20. Liu, X. et al. (2019). An adaptive CU size decision algorithm for HEVC intra prediction based on complexity classification using machine learning. *IEEE Transactions on Circuits and Systems for Video Technology (TCSVT), 29*(1), 144–155.

21. Correa, G. et al. (2019). Online machine learning for fast coding unit decisions in HEVC. In *Data Compression Conference (DCC)* (pp. 564–564).

22. Kim, K. et al. (2014). MC complexity reduction for generalized P and B pictures in HEVC. *IEEE Transactions on Circuits and Systems for Video Technology (TCSVT), 24*(10), 1723–1728.

23. Gao, et al. (2015). Fast intra mode decision algorithm based on refinement in HEVC. In *IEEE International Symposium on Circuits and Systems (ISCAS)* (pp. 517–520).

24. Correa, G. et al. Encoding time control system for HEVC based on rate-distortion-complexity analysis. In *IEEE International Symposium on Circuits and Systems (ISCAS)* (pp. 1114–1117).

25. Deng, X., & Xu, M. (2017). Complexity control of HEVC for video conferencing. In *IEEE International Conference on Acoustics, Speech and Signal Processing (ICASSP)* (pp. 1552–1556).
26. Bjontegaard, G. (2001). Calculation of average PSNR differences between RD Curves, VCEG-M33, ITU-T SG16/Q6 VCEG, 13th VCEG Meeting: Austin, USA, April 2001.

Versatile Video Coding (VVC)

2

This chapter provides a high-level description of VVC, starting with basic video coding concepts in Sect. 2.1. Section 2.2 describes the VVC hybrid encoder structure. Section 2.3 presents the VVC frames organization and block partitioning. The novelties at the VVC encoding tools are presented in Sect. 2.4. Finally, Sect. 2.5 presents the VVC Common Test Conditions (CTC), which explain the methodology required to evaluate new algorithms proposed for VVC.

2.1 Basic Video Coding Concepts

A video sequence consists of a series of frames (i.e., static images) presented sequentially to the viewer at a given temporal rate. At least 24 frames per second (fps) is required to achieve a smooth motion perception [1]. However, recent video coding applications have introduced more demanding requirements, increasing the need for higher frame rates to allow a better user experience. Generally, applications can require frame rates of up to 60 and 120 fps for High Definition (HD) and UHD videos, respectively [1]. Besides, even higher frame rates are considered for more immersive video [2].

Each frame is represented by a two-dimensional matrix of pixels with horizontal and vertical dimensions, which defines the spatial resolution. The spatial resolution of a video sequence can be arbitrarily defined, but some predefined formats are often adopted by industry and supported on many devices, e.g., 1280×720 (720p), 1920×1080 (1080p), and 3840×2160 (2160p). The perceived video visual quality is highly related to the number of pixels in the image; consequently, the higher the spatial resolution, the better tends to be the visual quality, leading to a better user experience.

Each matrix of pixels in fact is represented as three matrixes of samples. The number of bits used to represent each sample is typically 8 or 10 bits in the current video sequences.

© The Author(s), under exclusive license to Springer Nature Switzerland AG 2022
M. Saldanha et al., *Versatile Video Coding (VVC)*, Synthesis Lectures on Engineering, Science, and Technology, https://doi.org/10.1007/978-3-031-11640-7_2

Then, each pixel is represented with 24 or 30 bits, depending on the bit depth of the video. When using 24 bits per pixel, 16.7 million of different colors can be represented. If 30 bits per pixel are used, then more than 1 billion of different colors can be represented, which is an important improvement. Again, more immersive videos tend to use more than 30 bits per pixel, allowing even more realistic experiences.

There are several color spaces to represent the image in the digital domain, such as RGB (Red, Green, and Blue), HSV (Hue, Saturation, and Value), and YCbCr (luminance–Y, blue chrominance–Cb, and red chrominance–Cr). In all cases, three matrixes of samples are used, one for each color component.

In video coding applications, the YCbCr color space is widely adopted since this color space allows an independent processing of luminance information, which is most important to the human visual system (HVS), and the chrominance information, which is less relevant to HVS. Then, this color space allows a subsampling of chrominance information, which is commonly applied in consumer applications [1].

The subsampling of chrominance components is itself a technique of video coding since it discards part of the video data with minor or none perceptible visual degradation. The most common chrominance subsampling configurations adopted are 4:2:2 and 4:2:0. The 4:2:2 configuration encompasses two Cb and two Cr samples for four Y samples; whereas the 4:2:0 configuration includes one Cb and one Cr sample for four Y samples. The VVC design supports several types of chrominance subsampling. Since the chrominance subsampling configuration 4:2:0 is widely used for several video applications and most works in the literature, we have considered 4:2:0 subsampling configuration in all experiments carried out in this book.

2.2 VVC: A Hybrid Video Encoder

VVC was designed with a block-based hybrid video coding approach, an underlying concept of all major video coding standards, such as AVC and HEVC. As in other hybrid encoders, VVC splits each frame of the video sequence into blocks which are independently processed. These blocks are processed by intra- or inter-frame predictions, transform, quantization, entropy coding, inverse quantization, inverse transform, and filters. These encoder steps are common to other current video encoders. The use of this sequence of operations aims to reduce the amount of redundant data to represent the video information [3]. The spatial redundancy refers to the correlations between samples within a video frame [4]. The redundancies between neighboring frames are called temporal redundancies [5]. Finally, the entropic redundancy refers to the occurrence probabilities of the coded symbols.

Figure 2.1 presents the VVC encoder structure in a high level of abstraction. Considering this high level of abstraction, this figure could also represent other recent hybrid

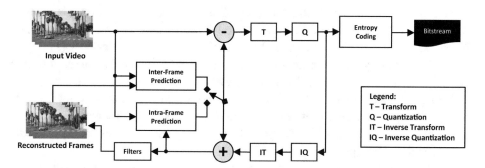

Fig. 2.1   General encoder structure

encoders, like AVC and HEVC. Figure 2.1 highlights the encoder data flow from the input video sequence to the generated output bitstream, including the main encoding steps.

An intra- or inter-frame prediction can be applied for each block. The intra-frame prediction explores the spatial redundancies inside the frame. In this case, only the video samples previously processed within the frame being encoded are used. The inter-frame prediction removes temporal redundancies using the information present in one or more frames previously encoded. If the information of only one frame is used to predict a block, then this is called unidirectional prediction. On the other hand, when the information of two frames is used to predict a block, then the prediction is called bidirectional, or simply bi-prediction.

The differences between the predicted and the original blocks are called residues. These residues are then processed by the transform and quantization steps, which are used to remove the spatial redundancies over the frequency domain. Firstly, the transform converts the residues from the spatial domain to the frequency domain. Thus, the quantization is applied over the transform coefficients. Since the HVS is less sensitive to high frequencies in an image, then, quantization discards or attenuates those frequencies that are less relevant to the HVS. Quantization is a lossy operation, since the discarded information cannot be recovered, but the level of losses can be controlled through a quantization parameter (QP). As higher is the QP value, as higher is the compression rate, but also as higher are the quality losses [1].

Finally, the entropy coding processes the quantized coefficients to reduce the entropic redundancy. In this case, the probabilities of occurrence of the values are explored and the values with a higher probability of occurrence are represented using codes with a smaller number of bits in the bitstream.

The inverse quantization and inverse transform are typical decoder operations, but they are also necessary inside the encoder to ensure that both encoder and decoder are using the same references. Afterward the inverse quantization and inverse transform, the reconstructed residual values are added to the predicted samples, generating reconstructed samples. These values are used as references for next intra-frame prediction operations, but to be used as a reference to the inter-frame prediction, firstly these samples must be filtered. The filter step smooths the artifacts inherent to the coding process for the reconstructed frames, increasing the visual quality and also increasing the coding efficiency [1].

VVC, as other hybrid encoders, uses a complex Rate-Distortion Optimization (RDO) process to define the encoder decisions, like the best block size, the best prediction mode, and other decisions. A dense set of encoding possibilities are evaluated to select the combination that provides the lowest Rate-Distortion cost (RD-cost). RD-cost is calculated based on the bit rate required for the prediction and the distortion between the predicted and original blocks [6].

VVC follows the same hybrid structure of previous encoders, like AVC and HEVC; however, VVC introduces several novel techniques and enhancements for block partitioning, intra- and inter-frame predictions, transform, quantization, entropy coding, and in-loop filters to improve the encoding efficiency. These techniques and enhancements are described in the following sections.

2.3   VVC Frames Organization and Block Partitioning

VVC organizes the frames in groups using a structure called Group of Pictures (GOP), containing all the required frame-decoding information. A GOP does not contain any data dependency with frames that do not belong to it. All GOPs in a video sequence have the same size, which is statically defined by an encoder configuration parameter.

A GOP can contain three types of frames: (i) I-frame, which is encoded using only intra-frame prediction; (ii) P-frame, which in addition to intra-frame prediction, allows inter-frame prediction from one reference frame per block using one motion vector and one reference index; and (iii) B-frame, which in addition to intra- and uni-prediction, also allows inter-frame prediction using two motion vectors and two reference indices [7].

Each video sequence frame can be divided into slices, which represent a frame region that does not have data dependencies from other regions inside the same frame. Each slice groups sequences of Coding Tree Units (CTUs), which can be divided into Coding Units (CUs). Like GOP, each CU in a slice is encoded according to a predefined slice type, which can be I, P, or B. Slices are used to avoid significant encoded data losses in case of bitstream transmission errors and are also used as a strategy for parallel processing since regions can be encoded/decoded without data dependencies. VVC also allows other frame divisions for different applications, such as Tiles, Wavefront, and Subpictures, which are better discussed in [8, 9].

Block partitioning plays an essential role in the compression efficiency of current video encoders. Therefore, the VVC experts investigated several new schemes of block partitioning structure to support block sizes larger than HEVC, providing an efficient compression rate, especially for high and ultra-high video resolutions. The VVC standard splits each input frame into CTUs covering square regions of at most 128×128 pixels. Each CTU is

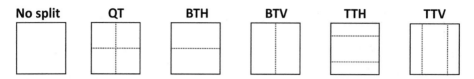

Fig. 2.2 QTMT partitioning possibilities

composed of one Coding Tree Block (CTB) of luminance and two CTBs of chrominance, one for Cb and the other for Cr. The size of the chrominance CTBs depends on the used color subsampling rate. If a subsampling rate of 4:2:0 is used, then a 128×128 pixels CTU will be composed of a luminance CTB with 128×128 samples, and two chrominance CTBs with 64×64 samples. Each CTU can be recursively partitioned into smaller blocks of pixels referred to as CUs. A CU is composed of a Coding Block (CB) of luminance samples and two CBs of chrominance samples. Again, the used subsampling rate is applied over the chrominance CBs, as in the chrominance CTBs.

VVC adopts a coding-tree-based splitting process that, in addition to the HEVC Quadtree (QT) structure, introduces the Multi-Type Tree (MTT) partitioning structure, enabling rectangular CU shapes through Binary Tree (BT) and Ternary Tree (TT). The combination of QT and MTT structures is named Quadtree with nested Multi-Type Tree (QTMT), allowing the six partitions in Fig. 2.2. A CU can be defined as no split, and the coding process is carried out with the current CU size. Otherwise, a CU can be split with QT, BT, and TT structures. QT splits a CU into four equal-sized square sub-CUs. BT splits horizontally (BTH) or vertically (BTV) a CU into two symmetric sub-CUs. TT splits a CU into three sub-CUs with the ratio of 1:2:1, and the split also can be performed in horizontal (TTH) and vertical (TTV) directions.

Firstly, a CTB is recursively partitioned with a QT structure. Subsequently, each QT leaf can be recursively partitioned with an MTT structure using binary and ternary splits. However, when an MTT split is performed, a QT split is no longer allowed. The CB sizes may vary from 128×128 (maximum CTB size) to a minimum of 4×4 samples, totalizing 28 block sizes possibilities, including all square and rectangular shapes. The intra-frame prediction supports a maximum CB size of 64×64 samples, then when a 128×128 luminance CTB is processed, this CTB will ever have a first QT partition, resulting in four 64×64 CBs. The QTMT leaves represent the CBs, which also are the units used for prediction and transformation.

Figure 2.3 exemplifies a QTMT partitioning structure of a 128×128 luminance CTB. The CTB is split into several CB sizes with different QT and MTT levels. Each colored line denotes a block partition type; black lines denote the QT splitting, gray and orange lines indicate the BTH and BTV splitting, respectively, and blue and purple lines represent TTH and TTV splitting, respectively.

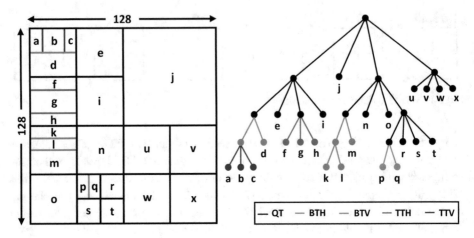

Fig. 2.3 Example of the QTMT CTB partitioning structure

Figure 2.4 presents a frame of a real video sequence superposed with the reached partition of the luminance CBs. The presented frame is the first frame of the BasketballPass video sequence encoded with VTM 10.0 [9] using all-intra configuration and QP 37.

As expected, the block partition structure is highly correlated with the image details. This conclusion can be noticed in Fig. 2.4a and b, where a detailed region is encoded with smaller CBs and a smooth region is encoded with larger CBs. Several MTT structure levels and different directions of BT and TT partitions are employed in this example of Fig. 2.4a (i.e., the detailed region), according to the texture characteristics. In contrast, few QT and MTT splitting levels are required to provide effective compression in the example of Fig. 2.4b (i.e., the smooth region).

VVC allows the use of one or two QTMTs to represent the CTU. If only one QTMT is used (a single tree), then the tree defined for the luminance CTB is applied for Cb and Cr CTBs (naturally considering the subsampling rate). If two QTMTs are used, then one QTMT is defined for the luminance CTB and other for the Cb and Cr CTBs (the same tree is used for both chrominance CTBs). This is called Dual-Tree (DT) or Chroma Separate Tree (CST). The single tree is employed for P- and B-frames (and for P- and B-slices), where both intra- and inter-frame predictions can be applied. The dual-tree is used only for I-frames (and I-slices), where only the intra-frame prediction can be performed. The CST was defined considering the finer texture granularity represented by luminance CTBs than the one represented by chrominance CTBs, which results in a higher number of smaller CBs in luminance partition than those in chrominance partition.

The QTMT partitioning structure provides higher flexibility for the encoder to represent the block sizes and shapes. Thus, these block partition types can adapt to various video characteristics resulting in higher coding efficiency. However, this high flexibility also results in a highest computational effort since the split possibilities are evaluated in

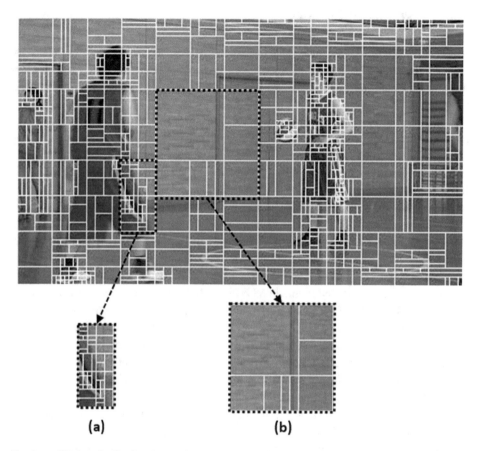

Fig. 2.4 CU size distribution for the first frame of BasketballPass (QP—37 and all-intra configuration): **a** detailed and **b** smooth regions

the RDO process to select the optimal CU partitioning. The optimal evaluation is performed recursively with all possibilities of splitting structures, including no split, QT, BTH, BTV, TTH, and TTV; however, simplifications are required due to its prohibitive computational effort. Some possible techniques are presented in Chaps. 6, 7, and 10 of this book.

2.4   VVC Encoding Tools

This section presents the main novelties of VVC considering the encoding tools. The discussion was divided into three parts. The first one presents the intra- and inter-frame prediction tools. The second one presents the VVC residual coding tools and the VVC entropy coding. Finally, the last one presents the VVC in-loop filters.

2.4.1 VVC Prediction Tools

VVC improves the intra- and inter-frame predictions, enhancing the HEVC-based coding tools and adopting novel approaches. This subsection briefly describes these improvements, and an extensive discussion of the intra-frame prediction is provided in Chap. 3.

Regarding the conventional intra-frame prediction applied in HEVC, VVC uses the Planar and DC modes like HEVC and provides finer-granularity angular prediction, extending the HEVC angular prediction modes from 33 to 65 angular modes. Since VVC allows the intra-frame prediction for rectangular block shapes, a specialized tool was developed to handle the angular prediction of these blocks: the **Wide-Angle Intra-Prediction (WAIP)** approach [10]. Moreover, intra-frame prediction evaluates **Position Dependent Prediction Combination (PDPC)** [11] and two types of 4-tap interpolation and smoothing filters to reduce errors and improve the coding efficiency. **Multiple Reference Line (MRL)** [12] allows more reference lines for VVC intra-frame prediction.

Matrix-based Intra-Prediction (MIP) [13] is a novel intra-frame prediction tool designed by a learning-based method and it is an alternative to the conventional angular intra-modes. **Intra-Sub-Partition (ISP)** [14] divides a CU into sub-partitions, reducing the distance of the reference samples, but all partitions must use the same coding mode.

VVC also brings a novel technique for chrominance intra-coding, called **Cross-Component Linear Model (CCLM)** [15]. CCLM predicts the chrominance samples based on the reconstructed luminance samples by applying linear models whose parameters are derived from reconstructed luminance and chrominance samples.

VVC maintains and enhances several HEVC inter-frame prediction tools, including Advanced Motion Vector Prediction (AMVP) [16] and merge mode, and introduces novel techniques. In VVC, the **Merge Candidate List (MCL)** construction process follows a specific order, and it is finished when all MCL positions are filled, with the maximum allowed length of six. This list is extended with five types of candidates in the following order:

(i)   Motion Vector (MV) from neighboring spatial CUs—at most four candidate blocks are selected from five possible spatial neighbors [9];

(ii) MV from neighboring temporal CUs—one candidate block is selected from two possible temporal neighbors [9];

(iii) History-based MV—In History-based Motion Vector Prediction (HMVP) [17], the MVs of the last five encoded blocks are stored and updated using a first-in-first-out (FIFO) rule [8];

(iv) Pairwise average MV—candidates are obtained from the average of MVs in the first two positions of MCL;

(v)   Zero MV—zero MVs (0, 0) candidates are appended to the list to fill it up [18].

Besides the merge mode that derives the motion information from neighboring, historical, or zero motion information, VVC introduces **Merge Mode with Motion Vector Difference (MMVD)** [18] to refine it. In MMVD, a base MV is selected from one of the first two candidates in MCL and a Motion Vector Difference (MVD) represented by a direction and a distance is encoded as a refinement [18]. Four directions and two distance tables with eight entries are predefined in the encoder, which signals the selected merge candidate, direction, and distance. More details about these tables can be obtained in [18].

The HEVC motion-compensated prediction considers only the translational motion model, which poorly represents many kinds of motion in a real-world video, such as rotation, scaling (zoom in/out), and shearing. **Affine Motion Compensation (AMC)** [19] predicts non-translational motion and improves the motion-compensated prediction efficiency as a new VVC tool. AMC describes the CU motion using MVs of two control points located at the top-right and top-left corners of the CU (four-parameter model) or three control points located at the top-right, top-left, and bottom-right corners of the CU (six-parameter model). Simpler motions such as rotation and scaling are represented in the four-parameter model, while more complex motions, such as shearing, require the application of the six-parameter model. Like the conventional motion-compensated prediction, VVC supports an affine AMVP and an affine merge mode for efficient prediction and coding of affine motion parameters. In the affine merge mode, VVC derives the control point MVs of the current CU based on the motion information of neighboring CUs [19].

Geometric Partitioning Mode (GPM) [20] is another VVC inter-frame prediction innovation, enabling motion compensation on non-rectangular block partitions as an alternative to the merge mode. When GPM is selected, the current CU is split into two partitions by a geometrically located straight line. In this case, each partition is predicted with its motion information, and only the unidirectional prediction with the merge mode is allowed; that is, each partition has one MV and one reference frame index. GPM supports 64 different geometric partitions. After predicting each partition, the sample values are combined using a blending processing with adaptive weights along the geometric partition line.

An inter-frame prediction signal can be combined with an intra-frame prediction signal using **Combined Inter/Intra-picture Prediction (CIIP)** [21]. This prediction combines merge mode with planar mode using weighted averaging, where the weight value is derived based on whether above and left neighboring CUs are encoded using intra-frame prediction or not.

VVC supports MVs accuracy for luminance samples of 1/16 instead of 1/4 supported in HEVC, which leads to a better prediction. Moreover, VVC adopts a scheme of adaptive resolution of MVs called **Adaptive Motion Vector Resolution (AMVR)** [18], where MVs can be coded with different accuracy according to the prediction mode, as shown in Table 2.1.

Table 2.1 Accuracy of VVC supported by MVs	Mode	Accuracy
	Conventional AMVP	1/4, 1/2, 1, or 4
	Affine AMVP	1/16, 1/4, or 1

VVC introduces three techniques called **Bi-prediction with CU-level Weights (BCW)** [22], **Bi-directional Optical Flow (BDOF)** [23], and **Decoder-side MV Refinement (DMVR)** [24] to improve the HEVC bi-prediction coding efficiency.

HEVC computes the bi-prediction averaging two prediction signals obtained from two different reference frames and/or using two MVs. VVC extends the bi-prediction using the **BCW** technique with a weighted averaging between the two bi-prediction signals being performed, where a predefined set of weights are evaluated through the rate-distortion cost.

BDOF is applied to optimize the bi-prediction signal and is based on human optical flow, which assumes that the movements of the objects are smooth. For each 4×4 sub-block in a CU, an MV is computed to refine the motion information and minimize the displacement difference between the samples of the two reference frames for bi-prediction.

DMVR refines the bi-prediction motion of the conventional merge mode by using a bilateral search step without extra data in the bitstream. This consists of an integer sample MV offset search and a refinement process with a fractional sample MV offset.

2.4.2 VVC Residual Coding and Entropy Coding

VVC enables a larger transform block than HEVC, bringing higher coding efficiency for larger resolution videos, such as 1080p and 4 K. VVC supports transform block sizes of up to 64×64 samples instead of 32×32 samples of HEVC, including square and rectangular transform block sizes that range from 4×4 to 64×64. HEVC allowed only squared transform unities with a maximum size of 32×32.

Besides, VVC introduces the **Multiple Transform Selection (MTS)** [25, 26] encoding tool that improves the transform module by including Discrete Sine Transform VII (DST-VII) and Discrete Cosine Transform VIII (DCT-VIII) to better decorrelate the residues. Like HEVC, VVC applies DCT-II for horizontal and vertical directions. Still, when MTS is enabled, DST-VII and DCT-VIII can be combined, and separate transforms in horizontal and vertical directions can be applied. MTS is allowed only for block sizes smaller or equal to 32×32.

Specific in the intra-frame prediction, a secondary transform is evaluated to low-frequency coefficients of blocks that selected DCT-II as the primary transform to explore better the directionality characteristics of the intra-frame prediction residues. This secondary transform is the **Low-Frequency Non-Separable Transform (LFNST)** [26, 27].

Specific in the inter-frame prediction, **Sub-Block Transform (SBT)** [26] encoding tool allows to encode only a sub-partition of the block and skips the remaining partition. There are eight SBT partition modes in VVC with different configurations of splits and sizes. When SBT is applied, the transform block is either the half or quarter size of the residual block regarding horizontal or vertical direction, and the remaining part of the block is discarded.

The quantization module processes the residual information of the encoding block, attenuating or removing frequencies that are less perceptible to the human visual system, however, causing information losses. As previously discussed, the Quantization Parameter (QP) defines the quantization level, where smaller QP values perform less attenuation in these frequencies, preserving the image details, while higher QP values cause a higher attenuation in these frequencies leading to a higher compression rate at the cost of image quality losses [28]. The main novelties in this module, when compared to the HEVC, are the maximum QP value increase from 51 to 63 and **Dependent Quantization (DQ)**, which allows the use of a second scalar quantizer [29, 30]. Furthermore, **Joint Coding of Chroma Residual (JCCR)** [30] allows using a single residual block for both chrominance components (Cb and Cr) when they are similar to each other.

As HEVC, the VVC entropy coding uses **Context Adaptive Binary Arithmetic Coding (CABAC)** [31] that performs a lossless entropy compression at the quantization results and at the lateral information (motion vectors, prediction modes, etc.). VVC introduces improvements, such as the multi-hypothesis probability update model, separate residual coding for transformed blocks and blocks encoded with Transform Skip Mode (TSM), and context modeling for transform coefficients [30]. Two estimators are maintained by VVC for the probability estimation, each estimator is independently updated with different adaptation rates, which are pre-trained based on the statistics of the associated bins [30].

With the entropy coding results, VVC can calculate the RD-cost of all possibilities of combinations among block partitioning, encoding modes, and transform types, enabling the encoder to select the combination that leads to the best rate-distortion.

2.4.3 VVC In-Loop Filters

Before processing the in-loop filters, **Luma Mapping with Chroma Scaling (LMCS)** [32, 33] is responsible for adjusting the range of the luminance input sample values to improve the subjective and objective quality of the encoded video sequence. Then, VVC applies three filters in the frame reconstruction loop, which are processed in the following order: **Deblocking Filter (DF)**, **Sample Adaptive Offset (SAO)**, and **Adaptive Loop Filter (ALF)**.

The DF reduces the blocking artifacts caused by the independent processing of CTUs and CUs. VVC DF has some improvements in relation to the HEVC DF. VVC extended the DF to consider the new block structures and coding tools with longer deblocking filters and a luminance-adaptive filtering mode designed specifically for HDR videos [32]. The SAO filter was inherited from HEVC and it is applied to attenuate the coding ringing artifacts, mainly caused by the quantization process. Finally, the ALF was adopted in some preliminary HEVC versions, but it was removed from the final standard version due to its high computational complexity. Although ALF still requires a high complexity, especially for the decoder, it can increase the coding efficiency significantly; thus, VVC has adopted ALF after some implementation changes. ALF is applied to correct further the signal based on linear filtering and adaptive clipping [32].

2.5   VVC Common Test Conditions

The Common Test Conditions (CTC) for VVC experiments [34] were defined by the experts of JVET to conduct the experiments in a fair and well-defined environment. So, it facilitates the comparison of results between the different techniques and tools since all tools are evaluated in the same context. The video sequences specified by CTC contain several characteristics to provide a robust evaluation. Thus, CTC is regularly updated to provide an evaluation that approximates the real environment of a video encoder.

CTC defines four main test configurations to be used in the evaluations:

(i)   **All-Intra (AI)**—The all-intra configuration defines that all frames should be only encoded with intra-frame prediction, i.e., all frames are defined as I-frames.

(ii) **Low-Delay (LD)**—In the low-delay configuration, only the first frame of the encoded video sequence is I-frame and the remaining frames are defined as B-frames, allowing inter-frame prediction with one or two reference frames.

(iii) **Low-Delay P (LDP)**—Low delay P configuration is similar to LD configuration, but in this case, B-frames are not allowed and only P-frames are used, i.e., the inter-frame prediction can use only one reference frame.

(iv) **Random-Access (RA)**—The random-access configuration uses a hierarchical temporal structure of B-frames, where I- and B-frames are employed, and the coding process is performed through GOPs.

The encoder works with 10-bit-depth for all configurations to represent each sample in the YCbCr format. The video sequences defined for the experiments are divided into six classes, including video resolutions from 416×240 to 3840×2160 pixels, totalizing 22 video sequences. Classes A1 and A2 refer to six video sequences of UHD 4 K (3840×2160 resolution). Class B has five video sequences with 1920×1080 resolutions. Class C and D represent videos with 832×480 and 416×240 resolutions, respectively, each one with four video sequences. Finally, Class E indicates three video sequences with 1280×720 resolution. Moreover, each video sequence should be encoded with 22, 27, 32, and 37 QP values.

The specifications of CTC video sequences are listed in Table 2.2. Besides, some images of 11 CTC videos (2160p and 1080p) are presented in Fig. 2.5.

The coding efficiency in CTC experiments is measured through Bjontegaard Delta Bit Rate (BD-BR) and Bjontegaard Delta Peak Signal-to-Noise Ratio (BD-PSNR) [35] metrics. BD-BR or BD-rate is a metric that presents the percentual of bitrate reduction of one encoder implementation in relation to the anchor for the same objective video quality. The objective video quality is measured in terms of Peak Signal-to-Noise Ratio (PSNR) [36]. In contrast, BD-PSNR is a metric that presents the PSNR reduction, in decibels (dB), of one encoder implementation in relation to the anchor for the same bitrate. The BD-BR and BD-PSNR metrics are calculated by interpolating the resulting bitrate and PSNR reached when encoding the novel and the anchor encoders considering the four QP values defined by the CTCs.

Table 2.2 Specifications of the videos in CTC

Class	Video sequence	Frames	Frame rate	Bit depth
A1	Tango2	294	60	10
	FoodMarket4	300	60	10
	Campfire	300	30	10
A2	CatRobot	300	60	10
	DaylightRoad2	300	60	10
	ParkRunning3	300	50	10
B	MarketPlace	600	60	10
	RitualDance	600	60	10
	Cactus	500	50	8
	BasketballDrive	500	50	8
	BQTerrace	600	60	8
C	BasketballDrill	500	50	8
	BQMall	600	60	8
	PartyScene	500	50	8
	RaceHorsesC	300	30	8
D	BasketballPass	500	50	8
	BQSquare	600	60	8
	BlowingBubbles	500	50	8
	RaceHorses	300	30	8
E	FourPeople	600	60	8
	Johnny	600	60	8
	KristenAndSara	600	60	8

Fig. 2.5 Frames of some CTC videos. **a** Tango2; **b** FoodMarket4; **c** Campfire; **d** CatRobot; **e** DaylightRoad2; **f** ParkRunning3; **g** MarketPlace; **h** RitualDance; **i** Cactus; **j** BasketballDrive; **k** BQTerrace

References

1. Wien, M. (2015). High efficiency video coding. Coding tools and specification. In *Signals and communication technology* (Vol. 24). Springer.
2. Salmon, R., et al. Higher frame rates for more immersive video and television. In *2011 British Broadcasting Corporation*. Retrieved October, 2021, from http://www.bbc.co.uk/rd/publicati ons/whitepaper209.
3. Richardson, I. (2010). *The H.264 advanced video compression standard* (2nd ed.). Chichester: Wiley.
4. Lainema, J., et al. (2012). Intra coding of the HEVC standard. *IEEE Transactions on Circuits and Systems for Video Technology (TCSVT)*, 22(12), 1792–1801.

5. Ghanbari, M. (2003). Standard codecs: Image compression to advanced video coding. S.l.: Institution Electrical Engineers.

6. Sullivan, G., & Wiegand, T. (1998). Rate-distortion optimization for video compression. *IEEE Signal Processing Magazine, 15*(6), 74–90.

7. Sullivan, G., et al. (2012). Overview of the high efficiency video coding (HEVC) standard. *IEEE Transactions on Circuits and Systems for Video Technology (TCSVT), 22*(12), 1649–1668.

8. Bross, B., et al. (2021). Overview of the versatile video coding (VVC) standard and its applications. *IEEE Transactions on Circuits and Systems for Video Technology (TCSVT), 31*(10), 3736–3764.

9. Chen, J., Ye, Y., Kim, S. (2020). Algorithm description for versatile video coding and test model 10 (VTM 10). In *JVET 19th Meeting, JVET-S2002, Teleconference.*

10. Zhao, L., et al. (2019). Wide angular intra prediction for versatile video coding. In *Data Compression Conference (DCC)* (pp. 53–62).

11. Pfaff, J., et al. (2021) Intra prediction and mode coding in VVC. *IEEE Transactions on Circuits and Systems for Video Technology (TCSVT), 31*(10), 3834–3847.

12. Chang, Y., et al. (2019). Multiple reference line coding for most probable modes in intra prediction. In *Data Compression Conference (DCC)* (pp. 559–559).

13. Schafer, M., et al. (2019). An affine-linear intra prediction with complexity constraints. In *IEEE International Conference on Image Processing (ICIP)* (pp. 1089–1093).

14. De-Luxán-Hernández, S., et al. (2019). An intra subpartition coding mode for VVC. In *IEEE International Conference on Image Processing (ICIP)* (pp. 1203–1207).

15. Zhang, K., et al. (2018). Enhanced cross-component linear model for chroma intra-prediction in video coding. *IEEE Transactions on Image Processing (TIP), 27*(8), 3983–3997.

16. Lin, J., et al. (2013). Motion vector coding in the HEVC standard. *IEEE Journal of Selected Topics in Signal Processing (JSTSP), 7*(8), 957–968.

17. Zhang, L., et al. (2019). History-based motion vector prediction in versatile video coding. In IEEE Data Compression Conference (DCC) (pp. 43–52).

18. Chien, W., et al. (2021). Motion vector coding and block merging in the versatile video coding standard. *IEEE Transactions on Circuits and Systems for Video Technology (TCSVT), 31*(10), 3848–3861.

19. Zhang, K., et al. (2019). An improved framework of affine motion compensation in video coding. *IEEE Transactions on Image Processing (TIP), 28*(3).

20. Gao, H., et al. (2021). geometric partitioning mode in versatile video coding: algorithm review and analysis. *IEEE Transactions on Circuits and Systems for Video Technology (TCSVT), 31*(9), 3603–3617.

21. Chiang, et al. (2018). CE10.1: Combined and multi hypothesis prediction. In *11th meeting, document JVET-K0257JVET*. Ljubljana, SI.

22. Chen, C., et al. (2016). Generalized bi-prediction method for future video coding. In *Picture Coding Symposium (PCS)* (1–5).

23. Alshin, A., Alshina, E. (2016). Bi-directional optical flow for future video codec. In *Data Compression Conference (DCC)* (pp. 83–90).

24. Gao, H., et al. (2021). Decoder-side motion vector refinement in VVC: algorithm and hardware implementation considerations. *IEEE Transactions on Circuits and Systems for Video Technology (TCSVT), 31*(8), 3197–3211.

25. Zhao, X., et al. (2016). Enhanced multiple transform for video coding. In *Data Compression Conference (DCC)* (pp. 73–82).

26. Zhao, X., et al. (2021). Transform coding in the VVC standard. *IEEE Transactions on Circuits and Systems for Video Technology (TCSVT), 31*(10), 3878–3890.

27. Koo, M., et al. (2019). Low frequency non-separable transform (LFNST). In IEEE Picture Coding Symposium (PCS) (pp. 1–5).
28. Budagavi, M., Fuldseth, A., Bjontegaard, G. (2014). HEVC transform and quantization. In *High efficiency video coding (HEVC): Algorithms and architectures* (pp. 141–169). Springer.
29. Schwarz, H., et al. (2019). Improved quantization and transform coefficient coding for the emerging versatile video coding (VVC) standard. In *IEEE International Conference on Image Processing (ICIP)* (pp. 1183–1187).
30. Schwarz, H., et al. (2021). Quantization and entropy coding in the versatile video coding (VVC) standard. *IEEE Transactions on Circuits and Systems for Video Technology (TCSVT), 31*(10), 3891–3906.
31. Marpe, D., Schwarz, H., Wiegand, T. (2003). Context-based adaptive binary arithmetic coding in the H.264/AVC video compression standard. *IEEE Transactions on Circuits and Systems for Video Technology (TCSVT), 13*(7), 620–636.
32. Karczewicz, M., et al. (2021). VVC In-Loop Filters. *IEEE Transactions on Circuits and Systems for Video Technology (TCSVT), 31*(10), 3907–3925.
33. Lu, T., et al. (2020). Luma mapping with chroma scaling in versatile video coding. In *Data Compression Conference (DCC)* (pp. 193–202).
34. Bossen, F., et al. (2020). VTM common test conditions and software reference configurations for SDR video. In *JVET 20th Meeting, JVET-T2010*.
35. Bjontegaard, G. (2001). Calculation of average PSNR differences between RD Curves. In *VCEG-M33, ITU-T SG16/Q6 VCEG, 13th VCEG Meeting*. Austin, USA.
36. Richardson, I.E. (2004). *H.264/AVC and MPEG-4 video compression—video coding for next-generation multimedia*. Chichester, Wiley.

VVC Intra-frame Prediction

3

This chapter describes the VVC intra-frame prediction encoding flow, specifying the enhancements in the HEVC-based coding tools and the novel coding tools for the intra-frame prediction and transform processes.

VVC intra-frame prediction allows luminance CB sizes ranging from 4×4 to 64×64 samples, while in chrominance, the maximum and minimum CB sizes are 32×32 and 16 samples (8×2 or 4×4), respectively. Moreover, VVC chrominance CBs are allowed to have 16×2 and 32×2 samples, but chrominance CBs with sizes 4×2 and $2 \times H$ (where H is the height of the block) are not allowed, intending to reduce the prediction time [1]. For I-slices, the MTT partitioning structure can be performed only over 32×32 or smaller CBs.

The VVC intra-frame prediction encoding flow for luminance samples as implemented at the VVC Test Model (VTM) version 10.0 [2, 3] is displayed in Fig. 3.1. The encoder evaluates several encoding possibilities aiming to minimize the RD-cost. Thus, the encoding process selects the prediction mode with the lowest RD-cost. Rough Mode Decision (RMD) and Most Probable Modes (MPM) [4] were inherited from the HEVC reference software [5, 6] to build the Rate-Distortion list (RD-list) containing the promising candidates. In this case, only the modes inside the RD-list will be evaluated through the full RDO process. RMD performs a local evaluation and roughly estimates the encoding cost of each candidate mode instead of evaluating all encoding possibilities by their RD-cost using the full RDO, which involves more complex operations, implying in a prohibitive computational effort.

RMD estimates the encoding efficiency of each mode through the Sum of Absolute Transformed Differences (SATD) (between the original and predicted block samples). Then, the modes are ordered according to their SATD-based costs, and the best modes are inserted into the RD-list. In the next step, MPM gets the default modes (the most frequently used ones) and the modes previously used to encode the left and above neighbor

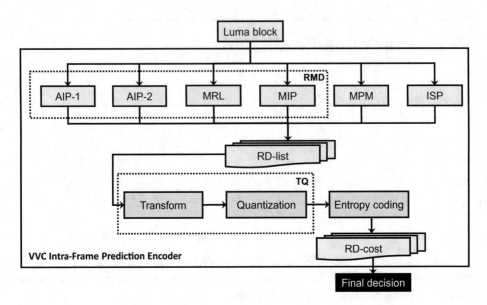

Fig. 3.1 VVC intra-frame prediction encoding flow for luminance blocks

blocks and inserts at most two additional modes into the RD-list. The RD-list starts with sizes of 8, 7, and 6 modes for 64×64, 32×32, and the remaining blocks (32×16, 16×32, 16×16, 32×8, 8×32, 32×4, 4×32, 16×8, 8×16, 8×8, 16×4, 4×16, 8×4, 4×8, and 4×4), respectively. However, the RD-list size can vary significantly according to the block size since it changes dynamically based on the encoding context and the use of fast decisions. At this point is important to highlight that without the RMD and the MPM heuristics, about 231 prediction modes should be evaluated by the full RDO process in the worst case, then the use of these heuristics brings an important impact in coding time reduction.

Angular Intra-prediction-1 (AIP-1) in Fig. 3.1 uses the HEVC intra-frame prediction modes, but VVC brings several novel intra-frame coding modes, including Angular Intra-prediction-2 (AIP-2) [7], Multiple Reference Line (MRL) [8], Matrix-based Intra-prediction (MIP) [9], Intra-Sub-Partition (ISP) [10], and an intra-mode coding method with six MPMs instead of three MPMs as in HEVC [11]. Planar mode is always coded first; then, DC and angular modes are coded using the remaining five positions of the MPMs list derived from intra-prediction modes from left and above neighboring blocks.

After evaluating these prediction tools, the modes inserted into the RD-list are processed by the residual coding, including the transform and quantization steps. The transform module encompasses Multiple Transform Selection (MTS) [12] and Low-Frequency Non-Separable Transform (LFNST) [13], as will be detailed in Sect. 3.6. The RD-costs are obtained after the transform and quantization steps by applying entropy coding.

For chrominance blocks, VVC intra-frame prediction inherits the HEVC prediction modes and inserts the Cross-Component Linear Model Prediction (CCLM) [14], where chrominance samples are predicted based on the reconstructed luminance samples by using a linear model.

Details of the novel intra-prediction tools defined in VVC are presented in the next sections.

3.1 Angular Intra-prediction

VVC extends the HEVC angular prediction modes from 33 to 65 angular modes to improve intra-frame prediction accuracy. All angular modes are illustrated in Fig. 3.2, where the solid gray lines depict the modes already used in HEVC intra-frame prediction, and dotted black lines are the ones introduced in VVC. Besides the angular modes, Planar and DC are also evaluated, resulting in 67 intra-frame prediction modes. These two modes remain with the same approach used in HEVC. Although Planar and DC are non-angular prediction modes, we call this tool Angular Intra-prediction (AIP) for simplicity.

AIP is divided into AIP-1 and AIP-2 in the VTM encoder as a fast mode decision, avoiding an exhaustive evaluation of the 67 intra-frame prediction modes for each block size. AIP-1 evaluates through the RMD process the Planar, DC, and 33 angular modes inherited from HEVC (solid gray lines in Fig. 3.2) and inserts a few modes into the RD-list. AIP-2 uses the RD-list information to detect the most promising new modes. It

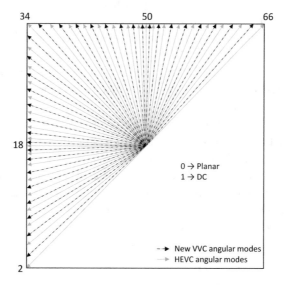

Fig. 3.2 VVC angular intra-frame prediction modes

evaluates the angular modes adjacent to the ones included in the RD-list (i.e., the best modes selected in AIP-1) and orders the RD-list based on the obtained SATD-based costs of these two steps. Thus, a reduced set of the new VVC angular intra-frame prediction modes is evaluated [2].

Wide-Angular Intra-prediction (WAIP) [15] is a new coding tool designed in VVC to handle rectangular block sizes that are now allowed in the QTMT partitioning structure for intra-frame prediction. This tool was developed because the conventional angular modes were designed in previous video encoding standards that allowed only squared blocks. Because of the predefined angles, rectangular blocks cannot reach good prediction samples. Thus, prediction modes with angles beyond 45° in the top-right direction are evaluated when the block width is larger than the block height. Otherwise, prediction modes with angles beyond 45° in the bottom-left direction are evaluated when the block height is larger than the block width. The number of evaluated intra-frame prediction modes is maintained since these wide-angle modes replace the prediction modes in the opposite direction with conventional angles [2]. Additionally, the DC prediction mode considers only the larger block side samples for rectangular blocks to provide a computationally efficient implementation.

3.2   Multiple Reference Line Prediction

VVC brings the possibility of using the MRL prediction [16], which allows more reference lines for the VVC intra-frame prediction than the single one used in HEVC. Figure 3.3 shows a block size of 4×4 samples and the reference lines used in the VVC intra-frame prediction when the MRL prediction is enabled.

Reference 1 (index 0) refers to the nearest reference line considered for the AIP tool. References 2 and 3 (indexes 1 and 2) are the two new reference lines that can be used to improve the coding efficiency of the intra-frame prediction since the adjacent reference line may significantly differ from the predicting block due to discontinuities, leading to a meaningful prediction error.

All combinations of prediction mode and reference line are evaluated using RMD, and MRL updates the RD-list that already contains the best modes selected in the AIP tool. However, due to the high computational cost required for evaluating all available intra-frame prediction modes with this extra number of reference lines, MRL requires simplifications. Thus, in VTM, the MRL prediction evaluates only MPMs, excluding Planar mode, for the two extra reference lines (lines 2 and 3) [11]. Planar mode is not considered for the MRL encoding tool since this combination does not provide additional coding gains for the VVC intra-frame prediction [11].

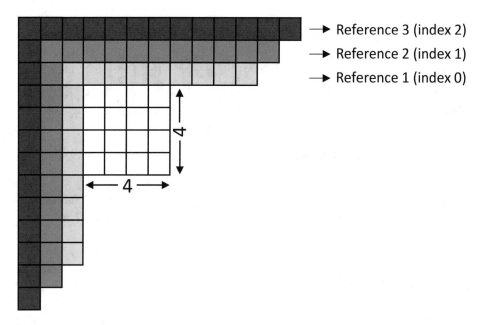

Fig. 3.3 MRL intra-frame prediction reference lines

3.3 Matrix-Based Intra-prediction

Matrix-based Intra-prediction (MIP) [9] is an alternative approach to the conventional angular intra-frame prediction modes, representing a new concept of intra-predictors designed by data-driven methods [11]. MIP is a machine learning-based algorithm where a set of matrices is defined according to the block size by offline training through neural networks, and each matrix represents a prediction mode.

The intra-frame prediction is performed using matrix multiplication and sample interpolation. An example of encoding an 8×8 samples CB is presented in Fig. 3.4, where neighboring samples of the adjacent reference lines are also used as prediction input. These neighboring samples are subsampled to perform the matrix multiplication, followed by the addition of an offset (b_k) and a linear interpolation (horizontal and vertical) to obtain the predicted block [9].

MIP defined 16 matrices for 4×4 blocks, eight for 8×8 blocks and blocks with exactly one side of length 4, and six for the remaining block sizes. VVC also allows using the transposed matrices, doubling the number of prediction modes. In VTM implementation, the predictions obtained in MIP are also evaluated applying RMD, and the RD-list is updated with the lowest SATD-based costs among AIP, MRL, and MIP prediction modes [2].

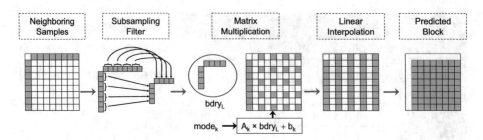

Fig. 3.4 Example of MIP flow for an 8 × 8 samples block

MIP can improve the encoding efficiency enabling predictions that vary in more than one direction (i.e., non-linear prediction), which is impossible with conventional angular modes.

3.4   Intra-sub-partition

Intra-Sub-Partition (ISP) explores the correlations inside a CB to improve the VVC intra-frame prediction [10]. The encoding CB is divided horizontally or vertically into sub-partitions which are sequentially encoded using the same intra-frame prediction mode. The sub-partitions are processed from top to bottom (horizontal split) or left to right (vertical split). The encoding sub-partition uses the reconstructed samples of previously encoded sub-partitions as a reference, increasing the reference sample correlation compared to the conventional approach, which only locates the reference samples at the left and the above boundaries of the predicting block. Figure 3.5 exemplifies two ISPs for a 64 × 64 CB.

Considering the 16 samples throughput, VVC defines some restrictions to the ISP mode: 4 × 8 and 8 × 4 CBs are split into only two sub-partitions (instead of four) and 4 × 4 CBs are not supported by the ISP. The remaining CB sizes can be split into four sub-partitions by ISP.

RMD cannot be applied together with the ISP since the real reconstructed samples must be used as a reference in the next sub-partition prediction, which can only be obtained by performing the complex RDO process. Consequently, VTM adopts some strategies

Fig. 3.5 Intra-Sub-Partition for a 64 × 64 CB split into 16 × 64 vertical and 64 × 16 horizontal sub-partitions

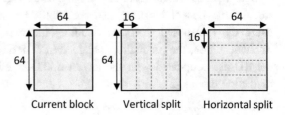

to derive the most promising prediction modes. ISP is applied only after the RDO is performed with the RD-list containing the best SATD-based costs among AIP, MRL, and MIP. Thus, ISP can use the information of the SATD-based costs and RD-costs of the AIP tool (i.e., AIP-1, AIP-2, and WAIP) to build a list of the most promising modes. The MRL and MIP tools are not considered for ISP mode derivation; then, the ISP list is generated alternating the split types (horizontal and vertical) in the following order:

(i)   Planar;
(ii) Angular modes ordered by RD-cost;
(iii) DC; and
(iv) The best AIP SATD-based costs, discarded from the RD-list after processing MRL and MIP.

These new prediction tools for luminance CBs improve the coding efficiency significantly, but this improvement comes at the cost of a high encoding effort since several prediction modes are evaluated in the complex RDO process. Some possible techniques to reduce this computational effort are presented in Chap. 8 of this book.

3.5 Encoding of Chrominance CBs

The VVC intra-frame prediction for chrominance CBs supports eight candidate modes: (i) Planar; (ii) Horizontal; (iii) Vertical; (iv) DC; (v) CCLM_LT; (vi) CCLM_L; (vii) CCLM_T; and (viii) Derived Mode (DM). The CCLM modes are based on a new tool called Cross-Component Linear Model, specially defined for chrominance samples.

The first four prediction modes (i.e., Planar, Horizontal, Vertical, and DC) are the same as those applied in luminance samples. Still, only the chrominance samples are considered for the prediction in this case.

The last four prediction modes (i.e., CCLM_LT, CCLM_L, CCLM_T, and DM) are inter-component prediction modes which explore the correlation between the luminance and chrominance components, where chrominance samples are predicted using information present in the luminance prediction and luminance reconstructed block.

The reconstructed luminance samples are used as references in the CCLM prediction modes to predict the chrominance samples, using a linear model as follows:

$$P(i, j) = a \cdot rec'_L(i, j) + b \tag{3.1}$$

where $P(i, j)$ refers to the predicted chrominance samples and $rec'_L(i, j)$ represents the reconstructed luminance samples. The reconstructed neighboring luminance and chrominance samples are used to derive the parameters a and b [14].

Three CCLM prediction modes are supported in VVC, including CCLM_LT, CCLM_L, and CCLM_T. Each prediction mode employs a different location of the reference samples used for model the parameter derivation. CCLM_L uses reference samples from the left boundary, CCLM_T involves reference samples from the top boundary, and CCLM_LT uses reference samples from the top and left boundaries.

Finally, DM refers to the prediction mode derived from the collocated luminance block; in other words, a prediction mode that has already been selected to encode the luminance block is inserted as a candidate to predict the chrominance block. When the derived mode is equal to any conventional prediction modes applied to luminance blocks (first four modes: planar, horizontal, vertical, or DC), mode 66 (diagonal 45°, see Fig. 3.2) is added in place of this conventional mode. When the collocated luminance block uses MIP, the Planar mode is applied, except for the 4:4:4 chrominance color format with a single partitioning tree where the same MIP mode is applied for the chrominance block [11].

3.6   Transform Coding

The intra-frame prediction transform coding has been improved in VVC by including Multiple Transform Selection (MTS) [12] and Low-Frequency Non-Separable Transform (LFNST) [13], which are tools for primary and secondary-transform modules, respectively. The dataflow model of the transform coding process for the VVC intra-frame prediction is displayed in Fig. 3.6 for luminance samples. The RD-list defined after applying the previous encoding algorithms is the input for the transforms, and VVC enables combining different transforms intending to minimize the RD-cost.

VVC brings two new transforms: (i) DST-VII and (ii) DCT-VIII, which can be used as an alternative to DCT-II, used as the main transform in HEVC. Using the best of these transform possibilities increase the contribution of this module to the global coding efficiency, but also increase a lot of the computational effort required to process the transforms, compared to HEVC. Moreover, the Transform Skip Mode (TSM) is available in VVC for 32×32 or smaller blocks, as in HEVC. In TSM, the prediction residues are directly sent to the quantization step, avoiding the use of transforms.

As displayed in Fig. 3.6, VVC transform coding has three encoding flows regarding the primary transform application:

(i)   the first flow using DCT-II for horizontal and vertical directions;
(ii) the second one using TSM; and
(iii) the third one using MTS, where DST-VII and DCT-VIII are used.

VTM processes these steps sequentially, starting with DCT-II, then TSM, and later processing MTS. The paths (i) and (ii) are similar to the HEVC transforms. The use of

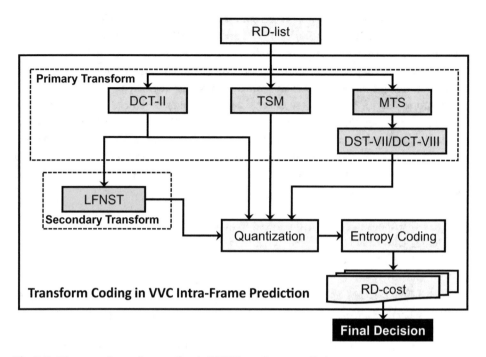

Fig. 3.6 Diagram of transform coding in VVC intra-frame prediction

MTS in the path (iii) allows joining DST-VII and DCT-VIII in horizontal and vertical directions; then, four combinations are evaluated [17]:

 (i) DST-VII and DST-VII;
 (ii) DST-VII and DCT-VIII;
 (iii) DCT-VIII and DST-VII; and
 (iv) DCT-VIII and DCT-VIII.

The primary transform block sizes are different where DCT-II has sizes ranging from 4×4 to 64×64, and DST-VII and DCT-VIII have sizes ranging from 4×4 to 32×32. Another important novelty is that these transforms support square and rectangular shapes. Finally, it is important to highlight that MTS is applied only for luminance samples.

The processing flow displayed in Fig. 3.6 has only one exception: when the CB is processed by the ISP tool, the combinations between DCT-II and DST-VII are also available. For other prediction tools, combinations of DCT-II and DST-VII/DCT-VIII are not allowed due to the limited coding gain and increased complexity of introducing additional encoding evaluations with more transform combinations [17].

High-frequency coefficients are zeroed out for transforming blocks with width or height equal to 64 for DCT-II and 32 for DCT-VIII and DST-VII to decrease the computational complexity. Thus, only low-frequency coefficients are retained.

LFNST is a non-separable secondary transform of the VVC intra-frame prediction that further decorrelates the low-frequency primary transform coefficients (top-left region of the transform block). LFNST may be applied for luminance blocks ranging from 4×4 to 64×64 samples that use DCT-II as a primary transform, as presented in Fig. 3.6, including square and rectangular block shapes. LFNST contains two secondary-transform sets (LFNST 1 and LFNST 2) with four non-separable transform matrices for each set [17]. The transform matrix evaluated for each set is defined based on the intra-frame prediction mode [17]. The VTM processes three cases:

(i)   DCT-II without LFNST (also referred to as LFNST 0);
(ii) DCT-II with LFNST 1;
(iii) DCT-II with LFNST 2.

When LFNST is not applied (i.e., case (i)), the DCT-II results are sent directly to quantization. Analogously to the luminance blocks, LFNST also can be applied for chrominance blocks.

These new transform coding tools provide significant improvements in the coding efficiency at the cost of a high computational effort, requiring efficient simplifications. Some possible techniques are presented in Chap. 9 of this book.

References

1. Huang, Y., et al. (2021). Block partitioning structure in the VVC standard. *IEEE Transactions on Circuits and Systems for Video Technology, 31*(10), 3818–3833.
2. Chen, J., Ye, Y., & Kim, S. (2020). Algorithm description for versatile video coding and test model 10 (VTM 10). In *JVET 19th Meeting, JVET-S2002, Teleconference.*
3. VTM. (2020). VVC Test Model (VTM). Retrieved October, 2021, from https://vcgit.hhi.fraunh ofer.de/jvet/VVCSoftware_VTM/-/releases/VTM-10.0.
4. Zhao, L., et al. (2011). Fast mode decision algorithm for intra prediction in HEVC. In *IEEE Visual Communications and Image Processing (VCIP)* (pp. 1–4).
5. HM. (2018). HEVC Test Model (HM). Retrieved October, 2021, from https://vcgit.hhi.fraunh ofer.de/jvet/HM/-/tags/HM-16.20.
6. Rosewarne, C., et al. (2015). High Efficiency Video Coding (HEVC) Test Model 16 (HM 16). In *Document: JCTVC-V1002.* Geneva, CH.
7. Bross, B., et al. (2021). Overview of the versatile video coding (VVC) standard and its applications. *IEEE Transactions on Circuits and Systems for Video Technology (TCSVT), 31*(10), 3736–3764.
8. Chang, Y., et al. (2019). Multiple reference line coding for most probable modes in intra prediction. In *Data Compression Conference (DCC)* (pp. 559–559).

9. Schafer, M., et al. (2019). An affine-linear intra prediction with complexity constraints. In *IEEE International Conference on Image Processing (ICIP)* (pp. 1089–1093).

10. De-Luxán-Hernández, S., et al. (2019). An intra subpartition coding mode for VVC. In *IEEE International Conference on Image Processing (ICIP)* (pp. 1203–1207).

11. Pfaff, J., et al. (2021). Intra prediction and mode coding in VVC. *IEEE Transactions on Circuits and Systems for Video Technology (TCSVT), 31*(10), 3834–3847.

12. Zhao, X., et al. (2016). Enhanced multiple transform for video coding. In *Data Compression Conference (DCC)* (pp. 73–82).

13. Koo, M., et al. (2019). Low frequency non-separable transform (LFNST). In *IEEE Picture Coding Symposium (PCS)* (pp. 1–5).

14. Zhang, K., et al. (2018). Enhanced cross-component linear model for chroma intra-prediction in video coding. *IEEE Transactions on Image Processing (TIP), 27*(8), 3983–3997.

15. Zhao, L., et al. (2019). Wide angular intra prediction for versatile video coding. In *Data Compression Conference (DCC)* (pp. 53–62).

16. Chiang, et al. (2018). CE10.1: Combined and multi hypothesis prediction. In *11th meeting, document JVET-K0257*. Ljubljana, SI: JVET.

17. Zhao, X., et al. (2021). Transform coding in the VVC standard. *IEEE Transactions on Circuits and Systems for Video Technology (TCSVT), 31*(10), 3878–3890.

State-of-the-Art Overview

4

This chapter presents the main works available in the literature proposing solutions to reduce the computational effort of the VVC intra-frame prediction. These works propose encoding time reduction solutions using heuristic and machine learning approaches focusing on different steps of intra-frame prediction, including block partitioning, intra-mode decision, and transform coding. It is important to mention that all works discussed in this chapter were evaluated under the all-intra encoder configuration.

The work of Fu et al. [1] presents a fast block partitioning technique using the Bayesian decision rule. A classifier is responsible for deciding when to skip the vertical split types based on the information derived from the evaluation of the current block (no split) and horizontal binary splitting. Besides, the horizontal ternary split evaluation is avoided when the cost of the vertical binary split is lower than the cost of the horizontal binary split. This technique was implemented in VTM 1.0 and reached a 45% encoding time reduction with a 1.02% increase in BD-BR [2].

Yang et al. [3] propose an encoding time reduction scheme encompassing a fast block partitioning solution based on decision tree classifiers and a fast intra-mode decision to decrease the number of angular intra-frame prediction modes evaluated. For this purpose, they trained one decision tree classifier for each split type (QT, BTH, BTV, TTH, and TTV), resulting in five classifiers. To decide the best split type before predicting the current block size, they used only texture information features of the current and neighboring blocks. The fast intra-mode decision applies a gradient descent search based on the MPMs to reduce the number of angular modes evaluated. This scheme was implemented in VTM 2.0 and the fast block partitioning solution reduces the encoding time by 52.59% with an increase of 1.56% in BD-BR, whereas the fast intra-mode decision saves 25.51% of the encoding time for a 0.54% BD-BR increase. The complete scheme provides a 62.46% encoding time reduction and an increase of 1.93% in BD-BR.

The work of Chen et al. [4] uses Support Vector Machine (SVM) classifiers to avoid the evaluation of horizontal or vertical split types in VVC intra-coding. For this purpose, they use six classifiers (one for each CU size, including 32×32, 32×16, 16×32, 16×16, 16×8, and 8×16) that are trained online using only texture information of the current block during the first frame encoding. The next frames are encoded by applying the decisions of the trained classifiers. This solution was implemented in VTM 2.1 and it reduces 50.97% of the encoding time with an increase of 1.55% in BD-BR.

Lei et al. [5] also propose a fast block partitioning solution to define the direction of BT and TT split types. This solution evaluates only a subset of angular intra-frame prediction modes considering virtual sub-partitions of the current block to estimate the horizontal and vertical splitting costs for the current block. Based on these costs, this solution can avoid horizontal or vertical splitting evaluations. This solution was implemented in VTM 3.0, reducing 45.8% of the encoding time with an increase of 1.03% in BD-BR.

Zhang et al. [6] present a scheme encompassing a solution to avoid the evaluation of some block partitions and a fast intra-mode decision to reduce the number of AIP modes evaluated in the intra-frame prediction. Regarding block partitioning, the solution applies a heuristic to classify the texture CU as homogeneous or complex. The texture complexity is measured based on the difference and standard deviation of the block samples, and the classification is carried out through a comparison with predefined threshold values. When a CU is classified as complex, Random Forest (RF) classifiers are used to define the split type to evaluate, including no split, QT, BTH, BTV, TTH, and TTV. When a CU is classified as homogenous, the splitting process is terminated, and no further partition types are evaluated. The fast intra-mode decision divides the AIP modes into four sets based on their directions, aiming to reduce the number of modes assessed in the RMD process. The Canny edge detector algorithm is used to identify the texture direction and to select only two sets of prediction modes to be evaluated. Then, the SATD costs of MPMs are computed, and the prediction modes with SATD costs higher than the MPMs are discarded, except Planar, DC, horizontal, and vertical modes. This scheme was implemented in VTM 4.0 and reduced 54.91% of the encoding time for a 0.93% BD-BR increase.

Tang et al. [7] also use the Canny edge detector algorithm, but in the block partitioning module. This algorithm is employed to identify the texture direction of the block and to classify the block texture complexity based on predefined threshold values. In this case, the texture orientation is identified to define the split types to be evaluated. When the block is classified as homogeneous, the splitting process is terminated; otherwise, the texture orientation is computed and classified as horizontal or vertical. When the algorithm classifies the texture as vertical, only BTV and TTV partitions are evaluated. In contrast, only BTH and TTH partitions are evaluated when the texture is classified as horizontal. The proposed solution provides a 36.18% encoding time reduction with a 0.71% BD-BR increase when implemented in VTM 4.0.1rc1.

The work of Cui et al. [8] proposes a fast block partitioning solution based on the direction of the sample gradients. This solution decides on three partitioning possibilities: split or not, horizontal or vertical, and BT or TT. The gradients of current block sub-partitions are computed in four directions and compared with predefined threshold values to decide on the best partitioning possibility. This solution was implemented in VTM 5.0, providing 50.01% encoding time saving at the cost of a 1.23% BD-BR increase.

The works [9–15] define the best block partitioning through encoding time reduction solutions based on Convolutional Neural Network (CNN). The solution [9] was implemented in VTM 6.1, and the experimental results demonstrated a 42.2% encoding time reduction at the cost of a 0.75% BD-BR increase. The solutions [10, 11] were implemented in VTM 7.0; the first solution reduced 39.39% of the encoding time with a 0.86% BD-BR increase and the second one saved 46.13% of the encoding time with a 1.32% BD-BR increase. The solution [12] was implemented in VTM 10.2 and the authors reported an encoding time reduction of 50% with a BD-BR increase of 0.7%. The solution [13] was implemented in VTM 10.0 and reduced 46.10% of the encoding time at the cost of a BD-BR increase of 1.86%. The solution [14] was implemented in VTM 7.0, and the experimental results showed a 39.16% encoding time reduction with a 0.7% BD-BR increase. The solution [15] was implemented in VTM 7.0 and reduces the encoding time by 45.81% with a BD-BR impact of 1.32%.

The work of Fan et al. [16] presents a fast block partitioning scheme based on the current block variance, sub-partition variances, and Sobel filter. The current block variance is calculated to verify the homogeneity of 32×32 blocks and early terminate the block partitioning evaluation. The sub-partition variances are computed to define only one split among QT, BTH, BTV, TTH, and TTV. Finally, the Sobel filter is used to decide on skipping MTT partitions and evaluating only the QT splitting. This scheme was implemented in VTM 7.0, reaching 49.27% encoding time reduction with a BD-BR increase of 1.63%.

Li et al. [17] propose a fast solution to skip BT and TT splitting based on residual block variances of sub-partitions, computed through the absolute difference between original and predicted samples. The absolute difference between variances of horizontal and vertical sub-partitions is calculated and compared with predefined threshold values to avoid unnecessary BT and TT evaluations. This solution was implemented in VTM 7.1 and it saves 43.9% of the encoding time at the cost of a 1.50% BD-BR increase.

The work of Zhang et al. [18] presents a strategy for computational effort reduction of arithmetic operations in the VVC transform coding process. This strategy explores the correlation among the coefficients of the DCT-VIII and DST-VII transform matrices to reduce the number of operations, reusing calculations for coefficients that are in flipped positions or that have only signal changes. This strategy was implemented in VTM 1.1 and evaluated with all-intra, random-access, and low-delay B configurations. The experimental results for all-intra configuration showed that this strategy reduces 3% of the encoding time without impacting the coding efficiency.

Fu et al. [19] propose an encoding time reduction scheme for the MTS coding tool in the intra-frame prediction. This scheme computes if all the neighboring spatial blocks were encoded with DCT-II to skip the DST-VII and DCT-VIII evaluations. Otherwise, the number of times that each transform is used in those blocks is verified and the transforms are evaluated from the most used to the least used. Additionally, if an intra-frame prediction mode in the current transform evaluation achieved a higher RD-cost than in the previously evaluated transform, that prediction mode is skipped in the next transform evaluations. This scheme was implemented in VTM 3.0 and it reduces 23% of the encoding time with a 0.16% BD-BR increase.

The work of Wu et al. [20] presents a fast block partitioning solution based on SVM classifiers. This solution employs one SVM classifier for each of the following binary decisions: (i) split or not and (ii) horizontal or vertical split. Considering the different CB sizes, the authors trained the classifiers separately for each size, resulting in a total of 16 classifiers. This solution was implemented in VTM 10.0 and provided a 63.16% encoding time reduction at the cost of a BD-BR increase of 2.71%.

The work [21] proposes a fast block partitioning scheme composed of two strategies. The first one employs a Bayesian decision rule using an estimated rate-distortion cost to decide when to finish the block partitioning evaluations. The second one uses an improved deblocking filter to predict the splitting mode according to the texture characteristics of the current CB. This scheme was implemented in VTM 11.0 and reduced 56.08% of the encoding time with a BD-BR increase of 1.30%.

Zhang et al. [22] present a fast scheme for VVC intra-frame prediction composed of a fast CB split decision and a fast intra-mode decision. The fast CB split decision uses the Bayesian decision rule to skip vertical and TT splits based on the information of previously encoded blocks. Besides, the Bayesian decision approach is also employed to finish the CB splitting process based on the information of RD-cost, QP, and variance of block samples. The fast intra-mode considers that the texture block is divided into four directions. It calculates the texture complexity to identify the texture direction of the CB and evaluates only the most promising angular mode along with planar and DC modes in the RDO process. This scheme was implemented in VTM 10.0 and provides a 59.52% encoding time reduction with an increase in BD-BR of 1.13%.

Liu et al. [23] developed a fast block partitioning algorithm based on the cross-block difference to avoid unnecessary horizontal and vertical evaluations of BT and TT partition structures. The cross-block difference is calculated for sub-blocks of the CB based on the Sobel filter for four directions, including horizontal, vertical, diagonal, and antidiagonal. The gradients are compared with predefined thresholds to decide when skipping a determined split type. This solution was implemented in VTM 9.3, reducing 41.64% of the encoding time with a BD-BR increase of 1.04%.

Song et al. [24] propose a fast block partitioning solution to decide when to avoid horizontal and vertical splitting. This solution measures the texture complexity through the mean absolution deviation of sub-blocks and compares it with predefined threshold values to estimate the texture direction of the CB. When the algorithm classifies the texture direction as horizontal, the vertical splitting evaluation is skipped. On the other hand, when the algorithm classifies as vertical, the horizontal splitting evaluation is skipped. This solution was implemented in VTM 6.3 and provides a 30.33% of encoding time reduction with a 0.57% BD-BR increase.

The work of Amna [25] presents a fast block partitioning solution to decide when to avoid TT-splitting evaluations. Firstly, this solution decides the TT direction based on the RD-cost obtained previously in BTH and BTV evaluations. Then, two Lightweight Neural Networks (LNN) are trained for each TT direction to determine if the split should be evaluated. The features used are based on the block size, MTT depth, transform coefficients, and RD-cost of previously encoded CBs. This solution was implemented in VTM 4.0 and reduces 46.91% of the encoding time with a BD-BR increase of 0.74%.

Zhao et al. [26] propose a fast block partitioning solution using a Deep Reinforcement Learning (DRL) approach. This solution is applied over 32×32 CBs and uses the RD-cost as a reward or a penalty for the DRL structure. The reward and penalty occur when the RD-cost decreases or increases with the decision of the network structure in the block partitioning evaluation. This solution was implemented in VTM 7.0 and reduces 54.38% of the encoding time with a BD-BR increase of 0.98%.

The work of Yao et al. [27] presents a fast mode decision algorithm using SVM to decide between planar and non-planar (DC and angular modes) intra-frame prediction modes. The features used as input in the SVM classifier are related to the prediction modes used in the neighboring CBs and statistical information of block samples, such as variance, Sobel filter, and a novel feature proposed by the authors called Statistical Oriented Gradient (SOG). This solution was implemented in VTM 5.0 and provides an 18.01% encoding time reduction with a 1.32% BD-BR increase.

Table 4.1 summarizes the works presented in this chapter. This table presents the VVC encoder module considered in each solution, including the partition structures QT and MTT, the mode decision, and the transform. The column "Mode" indicates that the work proposes a solution to reduce the computational effort of selecting intra-frame prediction modes, such as discarding the evaluation of some AIP modes in the RMD step. This table also provides the VTM version used for each work and experimental results in terms of BD-BR and Encoding Time Saving (ETS). Finally, the works in this table are sorted by the VTM version.

Table 4.1 Summary of the state-of-the-art intra-frame prediction

Work	QT	MTT	Mode	Transform	VTM	BD-BR (%)	ETS (%)
Fu et al. [1]		×			1.0	1.02	45.00
Zhang et al. [18]				×	1.1	0.00	3.00
Yang et al. [3]	×	×	×		2.0	1.93	62.46
Chen et al. [4]		×			2.1	1.55	50.97
Lei et al. [5]		×			3.0	1.03	45.80
Fu et al. [19]				×	3.0	0.16	23.00
Zhang et al. [6]	×	×	×		4.0	0.93	54.91
Amna et al. [25]		×			4.0	0.74	46.91
Tang et al. [7]	×	×			4.1	0.71	36.18
Cui et al. [8]	×	×			5.0	1.23	51.01
Yao et al. [27]			×		5.0	1.32	18.01
Tissier et al. [9]	×	×			6.1	0.75	42.20
Song et al. [24]		×			6.3	0.57	30.33
Zhao et al. [10]	×	×			7.0	0.86	39.39
Li et al. [11]	×	×			7.0	1.32	46.13
Fan et al. [16]	×	×			7.0	1.63	49.27
Huang et al. [14]	×	×			7.0	0.70	39.16
Javaid et al. [15]	×	×			7.0	1.32	45.81
Zhao et al. [26]	×	×			7.0	0.98	54.38
Li et al. [17]		×			7.1	1.50	43.90
Liu et al. [23]		×			9.3	1.04	41.64
Zhang et al. [13]	×	×			10.0	1.86	46.10
Wu et al. [20]	×	×			10.0	2.71	63.16
Zhang et al. [22]	×	×	×		10.0	1.13	59.52
Tech et al. [12]	×	×			10.2	0.70	50.00
Zhang et al. [21]	×	×			11.0	1.30	56.08

References

1. Fu, T., et al. (2019). Fast CU partitioning algorithm for H.266/VVC intra-frame coding. In *IEE International Conference on Multimedia and Expo (ICME)* (pp. 55–60).
2. Bjontegaard, G. (2001). Calculation of average PSNR differences between RD curves. In *VCEG-M33, ITU-T SG16/Q6 VCEG, 13th VCEG Meeting*. Austin, USA.

3. Yang, H., et al. (2020). Low complexity CTU partition structure decision and fast intra mode decision for versatile video coding. *IEEE Transactions on Circuits and Systems for Video Technology (TCSVT), 30*(6), 1668–1682.

4. Chen, F., et al. (2020). A fast CU size decision algorithm for VVC intra prediction based on support vector machine. *Multimedia Tools and Applications (MTAP), 79,* 27923–27939.

5. Lei, M., et al. (2019). Look-ahead prediction based coding unit size pruning for VVC intra coding. In *IEEE International Conference on Image Processing (ICIP)* (pp. 4120-4124).

6. Zhang, Q., et al. (2020). Fast CU partition and intra mode decision method for H.266/VVC. *IEEE Access, 8,* 117539–117550.

7. Tang, N., et al. (2019). Fast CTU partition decision algorithm for VVC intra and inter coding. In *IEEE Asia Pacific Conference on Circuits and Systems (APCCAS)* (pp. 361–364).

8. Cui, J., et al. (2020). Gradient-based early termination of CU partition in VVC intra coding. In *Data Compression Conference (DCC)* (pp. 103–112).

9. Tissier, A., et al. (2020). CNN oriented complexity reduction of VVC intra encoder. In *IEEE International Conference on Image Processing (ICIP)* (pp. 3139–3143).

10. Zhao, J., et al. (2020). *Adaptive CU split decision based on deep learning and multifeature fusion for H.266/VVC,* (Vol. 2020, pp. 1058–9244). Scientific Programming Hindawi.

11. Li, T., et al. (2021). DeepQTMT: A deep learning approach for fast QTMT-based CU partition of intra-mode VVC. *IEEE Transactions on Image Processing (TIP), 30,* 5377–5390.

12. Tech, G., et al. (2021). CNN-based parameter selection for fast VVC intra-picture encoding. In *IEEE International Conference on Image Processing (ICIP)* (pp. 2109–2113).

13. Zhang, Q., et al. (2021). Fast CU decision-making algorithm based on densenet network for VVC. *IEEE Access, 9,* 119289–119297.

14. Huang, Y.-H., Chen, J.-J., Tsai, Y.-H. (2021). Speed Up H.266/QTMT intra-coding based on predictions of resnet and random forest classifier. In *IEEE International Conference on Consumer Electronics (ICCE)* (pp. 1–6).

15. Javaid, S., et al. (2022). VVC/H.266 intra mode QTMT based CU partition using CNN. *IEEE Access, 10,* 37246–37256.

16. Fan, Y., et al. (2020). A fast QTMT partition decision strategy for VVC intra prediction. *IEEE Access, 8,* 107900–107911.

17. Li, Y., et al. (2021). Early intra CU size decision for versatile video coding based on a tunable decision model. *IEEE Transactions on Broadcasting (TBC), 67*(3), 710–720.

18. Zhang, Z., et al. (2019). Fast adaptive multiple transform for versatile video coding. In *IEEE Data Compression Conference (DCC)* (pp. 63–72).

19. Fu, T., et al. (2019). Two-stage fast multiple transform selection algorithm for VVC intra coding. In *IEEE International Conference on Multimedia and Expo (ICME)* (pp. 61–66).

20. Wu, G., et al. (2021). SVM based fast CU partitioning algorithm for VVC intra coding. In *IEEE International Symposium on Circuits and Systems (ISCAS)* (pp. 1–5).

21. Zhang, Q., et al. (2021). Fast CU partition decision method based on bayes and improved deblocking filter for H.266/VVC. *IEEE Access, 9,* 70382–70391.

22. Zhang, Q., et al. (2022). Low-complexity intra coding scheme based on Bayesian and L-BFGS for VVC. *Digital Signal Processing, 127,* 103539.

23. Liu, H., et al. (2021). Cross-block difference guided fast CU partition for VVC intra coding. In *International Conference on Visual Communications and Image Processing (VCIP)* (pp. 1–5).

24. Song, Y., et al. (2022). An efficient low-complexity block partition scheme for VVC intra coding. *Journal of Real-Time Image Processing (JRTIP), 19*(1), 161–172.

25. Amna, M., Imen, W., Fatma Ezahra, S. (2022). Fast multi-type tree partitioning for versatile video coding using machine learning. In *Signal, Image and Video Processing* (pp. 1–8).

26. Zhao, J., et al. (2022). Fast coding unit size decision based on deep reinforcement learning for versatile video coding. *Multimedia Tools and Applications, 81*(12), 16371–16387.

27. Yao, Y., et al. (2022). A support vector machine based fast planar prediction mode decision algorithm for versatile video coding. *Multimedia Tools and Applications, 81*(12), 17205–17222.

Performance Analysis of VVC Intra-frame Prediction

This chapter presents a performance analysis of VVC intra-frame prediction [1–3]. Section 5.1 details the methodology employed in the experiments. Section 5.2 compares VVC and HEVC regarding compression performance and computational effort. Section 5.3 analyzes the encoding time distribution of luminance and chrominance channels. Sections 5.4, 5.5, and 5.6 analyze the block size time and usage distribution, the encoding-mode time and usage distribution, and the transform time and usage distribution, respectively. Section 5.7 shows a rate-distortion and computational effort evaluation of VVC intra-frame coding tools. Lastly, Sect. 5.8 introduces a general discussion and indicates possibilities to reduce the computational effort of VVC intra-frame prediction.

5.1 Methodology

The analyses carried out in this chapter follow the test conditions presented in Sect. 2.3 with a 4:2:0 chrominance subsampling and all-intra configuration (only intra-frame prediction tools are used). The experiments were performed with VTM software version 10.0 [4, 5]. Even though new VTM versions are still being released with some new functionalities for different applications (e.g., high bit depth and other chrominance formats), the intra-coding flow for conventional camera-captured video sequence in VTM 10.0 remains without changes and achieves the same coding efficiency when compared to the latest VTM version.

The experiments considering the HEVC standard were performed in HM software version 16.20 [6], which is the HEVC reference software, employing all-intra configuration. For VTM and HM encoder configurations, the default temporal subsampling factor of eight was considered. In this case, the encoding process is carried out at every eight frames.

The computational effort and compression performance were computed through encoding time and BD-BR, respectively. Additional functions and changes were included in the VTM reference software to compute the encoding time and usage distribution of the encoding tools.

5.2   VVC Versus HEVC: Intra-frame Compression Performance and Computational Effort Evaluation

This section compares VTM and HM regarding BD-BR and encoding time. The experiments considered the video sequences defined in the VVC CTCs.

Figure 5.1 shows the BD-BR reduction of VTM when compared to HM for all-intra encoder configuration, considering the color channels of luminance (Y) and chrominance (Cb and Cr).

VTM significantly improves the compression efficiency for all classes of video sequences in both luminance and chrominance components. The highest BD-BR reductions for luminance and chrominance are noticed in class A2 with 29.3% and class A1 with 34.4%, respectively. These classes are the ones with the highest resolutions, as presented in Chap. 2. The smallest BD-BR reduction for luminance is in class D with 18.5%, and the highest BD-BR reductions occur for classes A1 and A2. This prominent gain for high-resolution video occurs especially due to the support for larger block sizes and due to the highly efficient block partitioning structure, which allows the encoding of larger blocks in uniform regions and flexible partition shapes in detailed regions. On average, VTM obtains a BD-BR reduction of 24.5, 23.6, and 24.5% for Y, Cb, and Cr, respectively. This analysis demonstrated that VVC provides a relevant higher compression rate than HEVC for intra-frame prediction, mainly for high-resolution videos.

Figure 5.2 presents the VTM encoding time increase over HM for the four QP values and each class of video sequences defined in CTC. QP 22 presents the highest increase in encoding time for all classes of video sequences, where VTM is 40 times slower than HM, on average. This result is expected since VTM carries out many block sizes and

Fig. 5.1 VTM compression efficiency improvement over HM in all-intra configuration

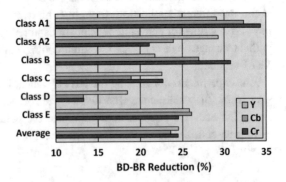

Fig. 5.2 VTM encoding time increase over HM in all-intra configuration

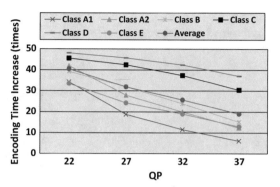

intra-frame prediction modes evaluations to preserve more image details for lower QPs. For QPs 27, 32, and 37, the VTM encoding time increases about 32, 26, and 19 times, respectively.

On the one hand, classes C and D present the highest encoding time increase (39 and 43 times, on average) because lower video resolutions tend to be encoded with smaller block sizes, implying the QTMT expansion to evaluate several block sizes and predict mode combinations. On the other hand, higher video resolutions tend to be encoded with larger block sizes and some fast-encoding decisions avoid the complete QTMT expansion. On average, VTM increases 27 times the encoding time over HM for intra-frame prediction, considering all classes and QPs.

One can conclude that the improvements in the VVC intra-frame prediction enable a higher compression rate than HEVC. Nevertheless, this high efficiency demands a high encoding effort, hindering real-time processing.

5.3 VVC Intra-frame Computational Effort Distribution of Luminance and Chrominance Channels

Figure 5.3 presents the VTM intra-frame encoding time distribution for luminance and chrominance components for the four QP values. VVC encoder allows the use of separate coding tree structures for luminance and chrominance blocks; consequently, this analysis aims to identify the computational effort required for each channel in the encoder.

The luminance coding flow requires the highest computational effort in the VTM intra-frame prediction. The minimum and maximum encoding time required for the luminance channel are 84 and 89% for QPs 22 and 32, respectively. On average, the luminance coding flow requires 87% of the VTM coding time in all-intra configuration. It occurs because the chrominance signal is subsampled, and the encoding flow evaluates fewer prediction tools. For instance, while luminance evaluates AIP [4], MRL [7], MIP [8],

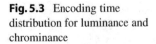

Fig. 5.3 Encoding time distribution for luminance and chrominance

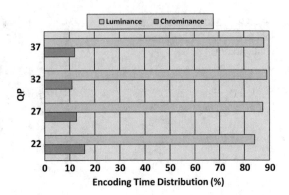

and ISP [9] for the prediction process and DCT-II/TSM, MTS [10], and LFNST [11] for the residual coding process, chrominance blocks evaluate only eight prediction modes applying DCT-II/TSM and LFNST for residual coding [12].

The subsequent analysis considers the encoding time and usage distribution of VVC intra-frame prediction focusing on analyzing the luminance block sizes and coding tools.

5.4 VVC Intra-frame Block Size Analysis

Figures 5.4 and 5.5 present the encoding time and usage distribution of luminance block sizes for the QP corner cases in CTC (22 and 37), respectively. These figures present the average results reached for all video sequences defined in the CTCs, for these two QPs. The x-axis, which varies from 64×64 to 4×4 samples, indicates the block sizes of both figures sorted from the largest to the smallest one allowed in the VVC intra-frame prediction. As discussed in Chap. 3, when processing I-slices [13], the MTT split is allowed for 32×32 or smaller blocks, then rectangular blocks with a width or height of 64 samples are not allowed. The red and gray bars in Figs. 5.4 and 5.5 indicate the results for QP 22 and 37, respectively.

The most time-consuming block sizes are 16×16, 16×8, or smaller for both QPs evaluated in Fig. 5.4. Nevertheless, the QP change causes different encoding time distributions for each block size. Higher QP values present a more heterogeneous encoding time distribution, whereas lower QP values concentrate the encoding effort in the block sizes with smaller areas. The total encoding time of block sizes smaller than or equal to 8×8 samples is 66.8% for QP 22 and 46.2% for QP 37.

From Fig. 5.5, it is possible to notice that the block size selection is highly dependent on the QP value. Higher and lower QP values imply using larger and smaller block sizes, respectively. Block sizes larger than 16×8 occur at less than 2% for QP 22, except for the 16×16, which occurs at 5.7%. For QP 22, the block usage concentrates in block sizes with 16×8 samples or smaller, representing 84% of the occurrences. In contrast, a more

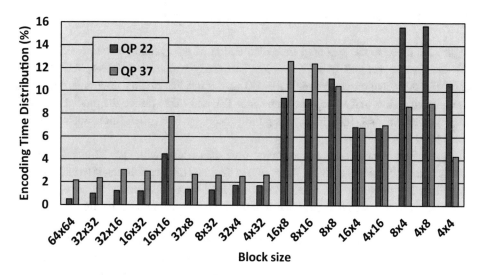

Fig. 5.4 Encoding time distribution of luminance block sizes for QPs 22 and 37

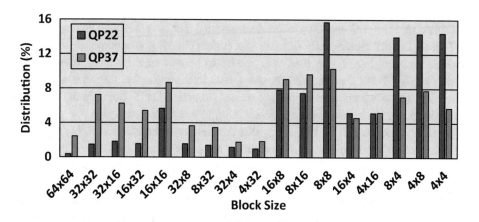

Fig. 5.5 Usage distribution of luminance block sizes for QPs 22 and 37

heterogeneous distribution is verified for QP 37, where the usage of block sizes larger than 16×8 and smaller than or equal to 16×8 samples are 40.7 and 59.3%, respectively. It occurs because low QP values retain more image details, producing more heterogeneous regions, which are better encoded with smaller block sizes. Oppositely, to raise the compression rate, high QP values attenuate the image details, producing more homogeneous regions that are better encoded with larger block sizes.

5.5   VVC Intra-frame Encoding Mode Analysis

Figure 5.6 presents the encoding time distribution of the encoding intra-prediction steps for each block size and QPs 22 and 37. This analysis considers AIP-1, AIP-2, MRL, and MIP as prediction steps and TQ + EC as the residual coding flow regarding transform, quantization, and entropy coding. Since ISP and MPM derive the prediction modes from predefined lists, they have negligible processing time; then, these tools were not considered in this analysis.

Figure 5.6a and b present the encoding time distribution with QPs 22 and 37, respectively. Since the residual coding (TQ + EC) is the most time-consuming step and the other steps together represent less than 30% of the total encoding time in all cases, Fig. 5.6 omits part of the residual coding distribution to allow a better visualization of the other steps.

One can notice that the computational effort of residual coding decreases for higher QPs; consequently, the prediction tools represent a higher percentage of the total encoding time. AIP-1 and MIP are the prediction tools that demand more encoding effort for both QPs. For QP 37, those tools achieve a maximum of 8.9 and 4.7% of the encoding time. MRL and AIP-2 demand, together, less than 4.5% of the encoding time, on average, in both QPs.

This high TQ + EC encoding effort demonstrated that novel residual coding tools like MTS and LFNST significantly increase the complexity of this module. Besides, VVC adds several new intra-frame prediction modes to be evaluated by the residual coding. Even though ISP has a negligible processing time in the prediction step, this coding tool can insert up to 48 prediction modes (16 modes for each LFNST index) in the RD-list to be evaluated by the residual coding, contributing to this high complexity.

Another conclusion from Fig. 5.6 is that the smaller is the block size, the higher the encoding effort spent in the prediction steps. It occurs mainly because of the relation between the available encoding tools and the number of samples per block size. This relation tends to concentrate the prediction effort in the smaller block sizes.

This analysis demonstrated that the residual coding of VVC intra-frame prediction had significantly raised its computational effort, achieving the highest encoding time for evaluated cases. It occurs because, for each prediction mode in the RD-list, the residual coding flow is executed several times, considering new prediction modes and different primary and secondary-transform combinations.

The following analysis presents the prediction mode selection distribution among the available intra-frame prediction modes. Figure 5.7 demonstrates this analysis considering AIP-1, AIP-2, MRL, MIP, and ISP prediction modes. For both QP values in Fig. 5.7a and b, AIP-1 is the most used mode, followed by MIP. MRL is more used than ISP for lower QPs, but the opposite happens for higher QPs. For both QPs, more than 45, 20, and 10% of the cases use AIP-1, MIP, and AIP-2, respectively.

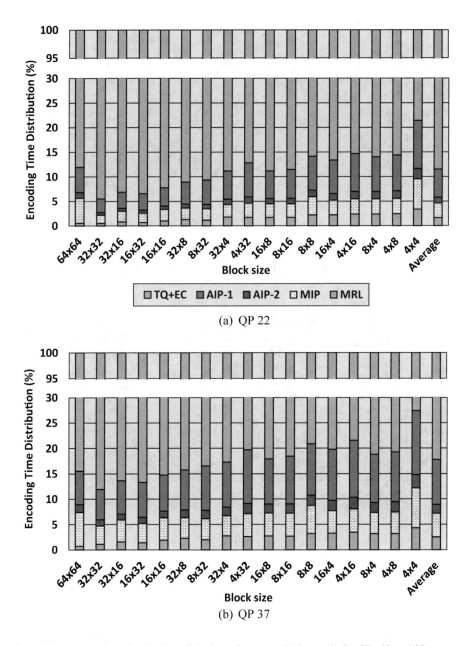

Fig. 5.6 Encoding time distribution of the intra-frame prediction tools for QPs 22 and 37

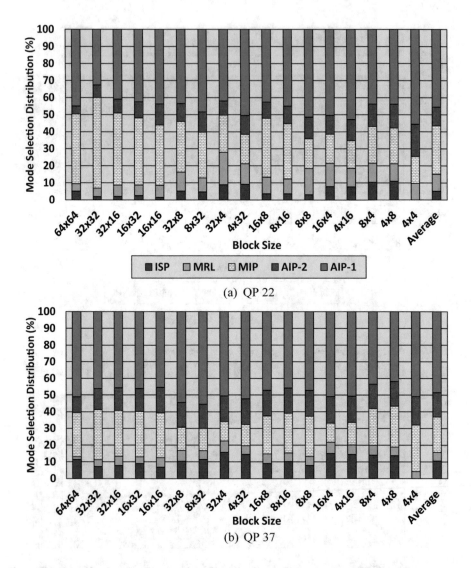

Fig. 5.7 Mode selection distribution of the intra-frame prediction tools for QPs 22 and 37

The QP value has a different impact on the encoding-mode distribution; the higher the QP value, the higher the use of AIP-1, AIP-2, and ISP tools. Hence, MIP and MRL present the opposite behavior. The block size also has a different impact on the encoding-mode usage; the higher the block size, the higher the use of MIP (mainly for lower QPs). Considering QP 22, MIP is even more used than AIP-1 for some blocks larger than $16 \times$

16 samples. AIP-2 and MRL tend to be less used for larger block sizes (mainly for lower QPs). ISP also follows this trend but with a less linear behavior.

This analysis showed that the HEVC intra-frame prediction modes (AIP-1) remain used a lot, providing high coding efficiency in many cases for VVC intra-frame prediction. However, the new VVC intra-frame coding tools also are crucial to raise the compression efficiency since these tools are used more than 51.5% of the times.

5.6 VVC Intra-frame Encoding Transform Analysis

Figure 5.8 shows the encoding time distribution of VVC primary transforms considering each block size and QPs 22 and 37. This analysis considers the following six horizontal and vertical transform combinations:

(i) DCT-II for both directions (DCT2_DCT2);
(ii) DST-VII for both directions (DST7_DST7);
(iii) DCT-II for horizontal and DST-VII for vertical direction (DCT2_DST7);
(iv) DST-VII for horizontal and DCT-II for vertical direction (DST7_DCT2);
(v) DST-VII for horizontal and DCT-VIII for vertical direction (DST7_DCT8);
(vi) DCT-VIII for horizontal and DST-VII for vertical direction (DCT8_DST7).

It is important to highlight that the encoding time of DCT2_DCT2 also encompasses the encoding time of TSM since VTM reference software assesses DCT2_DCT2 and TSM in the same execution flow (MTS index 0). Besides, as previously discussed in Sect. 3.2.6, the DCT-II and DST-VII transforms may be combined only for ISP-coded blocks with LFNST index 0 (without secondary transform). For this case, DST-VII is implicitly applied in the horizontal, vertical, or both directions if the block width, height, or both have between 4 and 16 samples (inclusive); otherwise, DCT-II is applied.

Although MTS enables DCT-VIII for both directions (DCT8_DCT8), our evaluations have not found this transform combination for any block size. Besides, combinations of DST-VII and DCT-VIII (DST7_DCT8 and DCT8_DST7) are low representative, with less than 0.1% of the encoding effort (on average), making it impossible to visualize them in Fig. 5.8.

These results demonstrate that DCT2_DCT2 is the most time-consuming transform operation for all block sizes and both QPs evaluated, followed by DST7_DST7 with the second-highest encoding effort. On average, DCT2_DCT2 and DST7_DST7 represent about 70% and 25% of the encoding effort, respectively. At the same time, the remaining transform combinations represent less than 6.8% of the encoding effort. For higher QPs, DCT2_DCT2 tends to require a slightly higher encoding effort than the other transforms.

DCT2_DCT2 encoding flow presents the highest encoding effort since this process in VTM evaluates three possibilities:

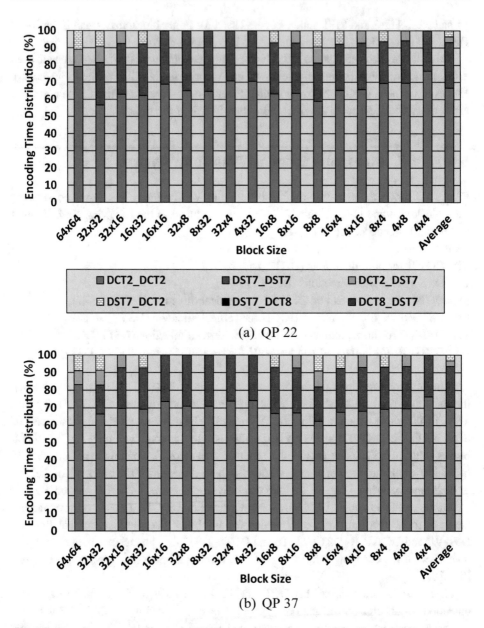

(a) QP 22

(b) QP 37

Fig. 5.8 Encoding time distribution of VVC primary transforms for QPs 22 and 37

 (i) DCT-II and TSM without secondary transform (i.e., LFNST index 0);
(ii) DCT-II with secondary-transform set one (LFNST index 1);
(iii) DCT-II with secondary-transform set two (LFNST index 2).

The remaining transform combinations do not perform LFNST; moreover, the following transform combination evaluations exclude the prediction modes that obtained high RD-cost applying DCT2_DCT2. Finally, VTM also implements fast decisions based on the obtained RD-cost by applying DCT-II/TSM to evaluate the next transform combinations conditionally.

Figure 5.9 displays the usage distribution of the transform combinations presented in Fig. 5.8. This analysis considers a TSM computation separated from DCT2_DCT2. As already expected, DCT2_DCT2 and DST7_DST7 are the most used transform combinations for all block sizes and QPs evaluated, representing more than 94% of the usage distribution, on average. For QP 37, DCT2_DCT2 is the most selected transform combination for all block sizes. However, for QP 22, DST7_DST7 is the most used transform combination for block sizes 32×16, 16×16, 16×8, 8×16, 16×4, and 4×16, indicating that the higher the QP value, the higher the selection of DCT2_DCT2 and the opposite behavior occurs for DST7_DST7.

Another observation is that the higher the QP value, the lower the use of TSM, DST7_DCT8, and DCT8_DST7 combinations. TSM is used 4.2 and 2.5% in QPs 22 and 37, respectively, and DST7_DCT8 and DCT8_DST7 are selected less than 0.1% in both cases. DCT2_DST7 and DST7_DCT2 have the opposite behavior, slightly increasing from 1.4 (QP 22) to 1.6% (QP 37); this occurs because these combinations are only employed for ISP prediction mode, which is more used for higher QPs. The transform behavior does not present any noticeable trend for most of the combinations regarding the block size. Only TSM has a clear trend to be more used for smaller block sizes for both QPs.

The low usage of transform combinations using DCT-VIII matrices occurs because MTS was designed without considering a secondary-transform operation for DCT-II. During the VVC standardization, LFNST was included in the encoder providing satisfactory rate-distortion performance for most cases by evaluating only the DCT-II (with and without LFNST), TSM, and DST-VII transforms.

Figure 5.10 shows the encoding time distribution for three LFNST encoding possibilities:

 (i) LFNST 0—only primary transform is applied.
(ii) LFNST 1—transform set one is applied.
(iii) LFNST 2—transform set two is applied.

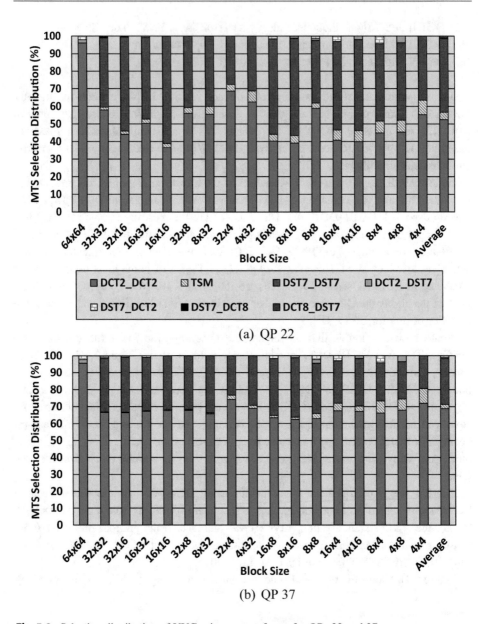

(a) QP 22

(b) QP 37

Fig. 5.9 Selection distribution of VVC primary transforms for QPs 22 and 37

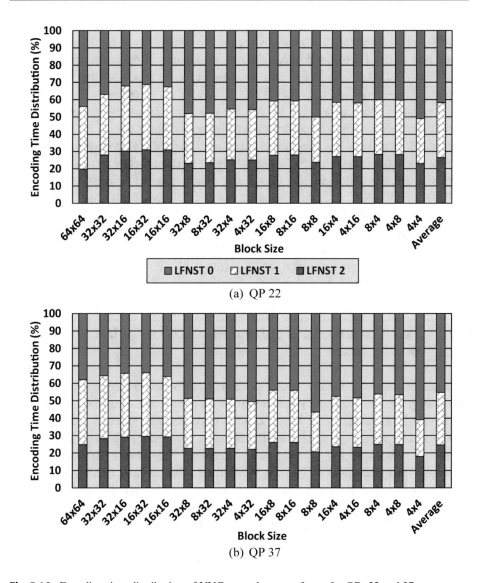

Fig. 5.10 Encoding time distribution of VVC secondary transforms for QPs 22 and 37

For both QPs, LFNST 0 presents the highest encoding effort, followed by LFNST 1 and LFNST 2. This occurs because the VTM encoder generates and processes the RD-list with DCT-II/TSM during the LFNST 0 evaluation. When LFNST 1 and LFNST 2 are processed, only DCT-II is evaluated, and the RD-list is derived from the LFNST 0 processing. Besides, the VTM encoder follows a sequential evaluation, where LFNST 0

is always evaluated. On the other side, LFNST 1 and LFNST 2 are conditionally evaluated based on the obtained RD-cost by performing LFNST 0 and the generated Coded
Block Flag (CBF) [4] of the previous evaluation, signaling if the block has any significant coefficients (i.e., non-zero). However, considering the total encoding time of both
secondary-transform evaluations (LFNST 1 and LFNST 2), it represents more than 55%
of the encoding effort, on average. Finally, one can conclude that the encoding effort of
LFNST does not directly correlate with the QP value and block size.

Figure 5.11 presents the selection distribution of the VVC secondary transforms for
all available block sizes and QPs 22 and 37. There is also no clear correlation with the
block size variation in this analysis. However, one can notice that the secondary transform
(LFNST 1 and LFNST 2) is more used for higher QP values, being used 55.3% of the
times with QP 37 and 29.1% of the times with QP 22, on average. This occurs because
LFNST is applied only for the DCT2_DCT2 transform combination; thus, this selection
distribution follows the same trend presented in the primary transform analysis, where
DCT2_DCT2 is also more used for QP 37. Finally, LFNST 1 is more used than LFNST
2 for all evaluated cases.

5.7   Rate-Distortion and Computational Effort of VVC Intra-frame Coding Tools

This section depicts a rate-distortion and encoding effort assessment of the new block
partitioning structure with binary and ternary splits and the novel intra-frame coding tools.
This analysis demonstrates the impact of each block partition structure and intra-frame
coding tool by removing it from the VTM encoding flow under all-intra configuration.

Table 5.1 shows the BD-BR increase and Encoding Time Saving (ETS) results when
removing binary partitions (BT), ternary partitions (TT), or both (MTT) of the intra-
frame encoding flow for luminance and chrominance blocks. When binary and ternary
partitions are removed, then the Multi-Type Tree (MTT) is completely removed and only
the Quadtree (QT) partitions are available. In other words, only squared CBs are allowed.

On average, when the BT partitioning is removed from the VTM encoder, the encoding
time is decreased by 77.1% at the cost of a 6.5% BD-BR increase. When removing the
TT partitioning, the encoding time is decreased by 48.4%, and BD-BR increases by 1.2%.
Removing both BT and TT partitions caused an encoding time reduction of 93.6%, with
a drawback in BD-BR of 26.1%, on average.

Even though the coding efficiency is reduced for all classes of test sequences, the
impact of removing BT and TT partitions is more prominent for video resolutions lower
than 3840×2160 (classes A1 and A2). It occurs because lower video resolutions are
better encoded with smaller block sizes, and BT and TT partition structures can enable
more block sizes and shapes, increasing the coding efficiency.

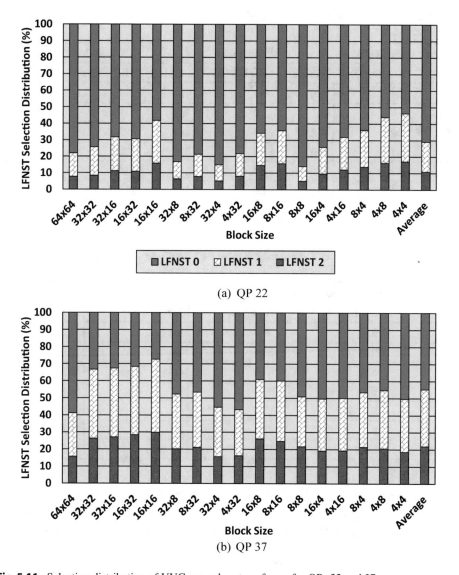

(a) QP 22

(b) QP 37

Fig. 5.11 Selection distribution of VVC secondary transforms for QPs 22 and 37

Table 5.2 presents the BD-BR and ETS results when removing horizontal (BTH and TTH) and vertical (BTV and TTV) partitions. This analysis shows that both vertical and horizontal partitions provide similar average results of ETS and BD-BR. The vertical and horizontal partition removal provides about 79.5% of ETS with a 5.5% BD-BR increase and 79.9% of ETS with a 5.3% BD-BR increase, respectively.

Table 5.1 Coding efficiency and encoding timesaving results when removing BT and TT partitioning structures

Class	Without BT		Without TT		Without MTT	
	BD-BR (%)	ETS (%)	BD-BR (%)	ETS (%)	BD-BR (%)	ETS (%)
A1	4.4	72.9	0.7	42.1	12.4	90.9
A2	4.9	78.7	1.0	48.6	16.0	94.5
B	5.8	77.6	1.1	48.4	22.1	94.7
C	8.5	79.7	1.6	51.9	36.2	95.5
D	6.7	77.5	1.3	51.4	30.7	93.1
E	8.6	76.0	1.8	48.1	39.3	92.9
Avg.	6.5	77.1	1.2	48.4	26.1	93.6

Table 5.2 Coding efficiency and encoding timesaving results when removing horizontal and vertical splits

Class	Without horizontal		Without vertical	
	BD-BR (%)	ETS (%)	BD-BR (%)	ETS (%)
A1	3.8	74.7	3.3	74.2
A2	3.7	80.4	4.0	80.0
B	5.6	81.0	4.3	80.0
C	6.5	83.2	6.5	82.6
D	5.7	81.9	6.2	80.9
E	6.6	78.4	8.6	79.0
Avg.	5.3	79.9	5.5	79.5

These results displayed similar behavior to the results presented in Table 5.1, showing the high coding efficiency provided by horizontal and vertical partitions for all test sequences and a higher BD-BR impact for lower video resolutions. Both analyses demonstrate that the new QTMT partitioning structure provides impressive coding efficiency at the cost of high computational effort.

Table 5.3 shows the BD-BR and ETS results when removing intra-frame coding tools of the VVC encoder. On the one hand, the highest encoding efficiency impacts are obtained when removing the residual coding tools LFNST and MTS with a 1.2% BD-BR increase. On the other hand, the highest ETS results are attained when removing LFNST or ISP tools.

Table 5.3 Coding efficiency and encoding timesaving results when removing VVC intra-frame coding tools

Class	Without AIP-2		Without MRL		Without MIP	
	BD-BR (%)	ETS (%)	BD-BR (%)	ETS (%)	BD-BR (%)	ETS (%)
A1	0.6	−0.7	0.1	0.8	1.0	10.3
A2	0.6	1.3	0.2	0.4	0.6	10.7
B	0.7	1.4	0.4	−0.1	0.5	11.4
C	0.9	1.6	0.7	0.8	0.5	13.0
D	0.8	0.4	0.2	0.9	0.6	11.7
E	1.4	0.2	0.3	−0.1	0.7	10.5
Avg.	0.8	0.7	0.3	0.5	0.6	11.3
Class	Without ISP		Without MTS		Without LFNST	
	BD-BR (%)	ETS (%)	BD-BR (%)	ETS (%)	BD-BR (%)	ETS (%)
A1	0.1	13.0	1.4	10.3	1.8	24.8
A2	0.3	14.2	1.4	13.9	0.7	28.7
B	0.4	15.1	1.4	14.3	1.0	26.7
C	0.7	17.7	0.9	15.7	1.4	27.1
D	0.6	15.9	0.7	15.2	1.1	25.4
E	0.8	15.1	1.4	14.0	1.5	25.5
Avg.	0.5	15.2	1.2	13.9	1.2	26.4

Removing AIP-2 and MRL provides few gains in ETS (less than 1%); removing MRL represents the smallest impact on the coding efficiency, and removing AIP-2, BD-BR is increased by almost 1%, presenting the highest BD-BR increase among the prediction tools. The MIP removal reduces 11.3% of ETS with a 0.6% BD-BR increase.

The highest BD-BR impacts of AIP-2, MRL, and MIP prediction tools are 1.4% (Class E), 0.7% (Class C), and 1.0% (Class A1), respectively. ISP presents the highest and lowest BD-BR impacts for classes E and A1, respectively. MTS and LFNST transform tools present similar behavior between the classes of video sequences, except in classes C and D, where the MTS coding tool obtained a lower BD-BR impact than the others.

This evaluation demonstrated that each new VVC intra-frame prediction tool improves the coding efficiency. Nevertheless, this improvement comes at the cost of a high encoding effort, mainly for the residual coding that runs many times to select the best combination of prediction mode, primary transform, and secondary transform.

5.8   General Discussion

Based on the analyses presented in previous sections, several ideas and conclusions can be taken to develop efficient timesaving solutions for the VVC intra-frame prediction. Firstly, small block sizes such as 4×4, 4×8, 8×4, and 8×8 require more encoding time compared to larger block sizes regardless of the QP value. Besides, the usage of these small block sizes decreases according to the QP increase. However, larger block sizes are less frequently used with low QP values. Thus, the quantization scenario can be considered to design efficient encoding timesaving solutions by adaptively avoiding the evaluation of some block sizes in the QTMT structure. Since the QTMT structure comprises three partitioning structures (QT, BT, and TT), some approaches for encoding timesaving solutions can be explored:

(i)   Predict the quadtree depth level, efficiently pruning this tree;
(ii) Predict the MTT depth level, efficiently pruning this tree;
(iii) Predict when avoiding BT and/or TT split evaluations;
(iv) Predict when avoiding horizontal and/or vertical split evaluations.

The analyses of the intra-frame coding flow showed that TQ + EC is the most time-demanding module regardless of the QP value. Hence, solutions to reduce the number of prediction modes evaluated in the TQ + EC flow (i.e., reduce the RD-list) must be explored to achieve more impressive encoding timesaving results. When considering the prediction steps, a more limited encoding timesaving can be achieved. In this case, AIP-1 is the most time-consuming step, and reducing the number of prediction modes evaluated in the RMD search can also decrease the encoding effort.

Besides, the residual coding is responsible for a noticeable encoding time to evaluate each intra-frame prediction mode through the TQ + EC flow for all available block sizes. This encoding effort is mainly due to the evaluations of primary and secondary transforms. For primary transform, DCT2_DCT2 and DST7_DST7 are the most time-consuming transform operations. DST7_DST7 is more used for low QP values (reducing the selection of DCT2_DCT2), whereas for high QP values, an opposite behavior occurs, decreasing the use of DST7_DST7 and increasing the use of DCT2_DCT2 significantly. For the secondary transform, the three possibilities of LFNST encoding demand similar encoding effort; nevertheless, LFNST 1 and LFNST 2 are less frequently used for lower QP values, whereas the opposite occurs for higher QP values. Therefore, solutions exploring the encoding context can be designed to reduce the transform combinations evaluations in the intra-frame coding flow, including predicting the primary transform combination and avoiding unnecessary secondary-transform evaluations.

References

1. Saldanha, M., et al. (2020). Complexity analysis of VVC intra coding. In *IEEE International Conference on Image Processing (ICIP)* (pp. 3119–3123).
2. Saldanha, M., et al. (2021). Analysis of VVC intra prediction block partitioning structure. In *Visual Communications and Image Processing (VCIP)* (pp. 1–5).
3. Saldanha, M., et al. (2021). Performance analysis of VVC intra coding. *Journal of Visual Communication and Image Representation (JCVIR), 79*, 103202.
4. Chen, J., Ye, Y., Kim, S. (2020). Algorithm description for versatile video coding and test model 10 (VTM 10). In *JVET 19th Meeting, JVET-S2002, Teleconference.*
5. VTM. (2020). VVC Test model (VTM). Retrieved October, 2021, from https://vcgit.hhi.fraunhofer.de/jvet/VVCSoftware_VTM/-/releases/VTM-10.0
6. Rosewarne, C., et al. (2015). High efficiency video coding (HEVC) test model 16 (HM 16). In *Document: JCTVC-V1002*, Geneva.
7. Chang, Y., et al. (2019). Multiple reference line coding for most probable modes in intra prediction. In *Data Compression Conference (DCC)* (pp. 559–559).
8. Schafer, M., et al. (2019). An affine-Linear intra prediction with complexity constraints. In *IEEE International Conference on Image Processing (ICIP)* (pp. 1089–1093).
9. De-Luxán-Hernández, S., et al. (2019). An intra subpartition coding mode for VVC. In *IEEE International Conference on Image Processing (ICIP)* (pp. 1203–1207).
10. Zhao, X., et al. (2016). Enhanced multiple transform for video coding. In *Data Compression Conference (DCC)* (pp. 73–82).
11. Koo, M., et al. (2019). Low frequency non-separable transform (LFNST). In *IEEE Picture Coding Symposium (PCS)* (pp. 1–5).
12. Pfaff, J., et al. (2021). Intra prediction and mode coding in VVC. *IEEE Transactions on Circuits and Systems for Video Technology (TCSVT), 31*(10), 3834–3847.
13. Huang, Y., et al. (2021). Block partitioning structure in the VVC standard. *IEEE Transactions on Circuits and Systems for Video Technology, 31*(10), 3818–3833.

Heuristic-Based Fast Multi-type Tree Decision Scheme for Luminance

6

This chapter presents a design of a fast decision scheme based on statistical analysis for MTT luminance-block partitioning structure [1]. When the intra-frame prediction is evaluated, AIP [2] and ISP [3] modes tend to indicate the texture direction of the encoding block. Then, the best mode selected by these tools can be an effective predictor for the MTT partitioning decision and be used as a fast mode decision. Moreover, characteristics can be extracted from the block samples to decide the direction of the binary and ternary split since MTT tends to divide the block into regions, sharing more similar sample values for providing accurate predictions. Therefore, this chapter presents the design of an accelerating scheme composed of two strategies exploring the correlation of the intra-frame prediction modes and samples of the current CB to decide the split direction of binary and ternary partitions. According to this decision, our scheme avoids unnecessary evaluations of binary and ternary partitions, reducing the encoding time with a minimum impact on the coding efficiency. This scheme was published at IEEE International Symposium on Circuits and Systems (ISCAS) [4].

6.1   Initial Analysis

Figure 6.1 displays the luminance CB partitions for the first frame of the BasketballPass video sequence, encoded with QP 37 and all-intra configuration. Along with the partition distribution, to support a visual analysis, the three blocks with horizontal and vertical splits are detached in blue and red boxes, respectively.

Two variance values were defined for each luminance block aiming to extract from the luminance samples the possible direction of binary or ternary splits:

© The Author(s), under exclusive license to Springer Nature Switzerland AG 2022
M. Saldanha et al., *Versatile Video Coding (VVC)*, Synthesis Lectures on Engineering, Science, and Technology, https://doi.org/10.1007/978-3-031-11640-7_6

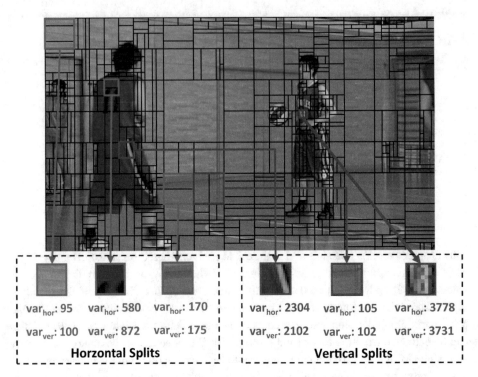

Fig. 6.1 CB partitioning of BasketballPass video sequence with variance values of the highlighted blocks

(i) var_{hor}–the sum of the variance values of upper and lower partitions, considering the current block is horizontally subdivided into two equal-sized regions.

(ii) var_{ver}–the sum of the variance values of the left and right partitions for a current block vertically subdivided into two equal-sized regions.

Figure 6.1 shows that var_{hor} tends to be smaller than var_{ver} when the horizontal CB partitioning (blue boxes) occurs; also, CB tends to split vertically when var_{ver} is smaller than var_{hor} (red boxes). These variances extracted from the encoding block can be analyzed to predict the binary and ternary split direction, skipping several unnecessary MTT evaluations.

Additionally, we performed a second preliminary analysis that correlates the encoding context through the AIP modes when it was chosen as the best mode for the current CB to further improve the performance of our scheme. Therefore, inspired by the work of Fu et al. [5], the AIP modes were divided into two categories:

(i) Horizontal directions (AIP_{hor})–AIP modes from 10 to 28 (horizontal mode ± 8).

(ii) Vertical directions (AIP_{ver})–AIP modes from 42 to 58 (vertical mode ± 8).

Fig. 6.2 Accuracy for FDV, FD-ISP, and overall strategies

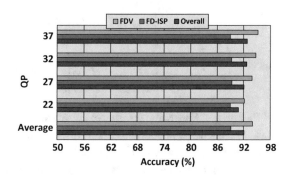

Considering this division, we proposed a strategy called Fast Decision based on Variance (FDV) to identify the texture direction according to the variance values and the best AIP mode of the current luminance CB. On the one hand, the encoding block should probably be horizontally split if var_{hor} is smaller than var_{ver} and the intra-frame prediction mode is AIP_{hor}; then, the encoder can skip the vertical binary and ternary splitting evaluations. On the other hand, if var_{ver} is smaller than var_{hor} and the intra-prediction mode is AIP_{ver}, the encoder can skip the horizontal binary and ternary splitting.

The second strategy, named Fast Decision based on ISP (FD-ISP), correlates ISP with the MTT structure due to the partitioning similarity. This strategy only allows the splitting in the same direction as the ISP mode selected. In other words, the encoder skips the vertical or horizontal binary and ternary splitting evaluations if the best mode is ISP_{hor} or ISP_{ver}, respectively.

Seven video sequences (Campfire, CatRobot, BasketballDrive, RitualDance, RaceHorsesC, BasketballPass, and FourPeople) were encoded in all-intra configuration with four QP values (22, 27, 32, and 37) to determine the accuracy of both strategies. The accuracy of FDV, FD-ISP, and Overall (both strategies jointly evaluated) are presented in Fig. 6.2. By accuracy, we mean the ratio of correct predictions divided by the total number of instances evaluated. One can notice that the FDV accuracy is higher than 92% for all cases assessed, whereas the FD-ISP accuracy is higher than 88%. The overall results demonstrate the proposed scheme's efficacy, with 92.1% accuracy, on average.

6.2 Designed Scheme

Equation 6.1 defines FD_{hor}, which is responsible for the fast decision to skip the horizontal binary/ternary splitting evaluations. Equation 6.2 specifies FD_{ver}, which controls the fast decision to skip the vertical binary/ternary splitting. If FD_{hor} is true, then the horizontal partitions are skipped in the MTT, including binary and ternary horizontal partitions. On the other hand, if FD_{ver} is true, then the vertical partitions are skipped, including binary and ternary vertical partitions.

$$FD_{hor} = \begin{cases} True, & if\ (AIP_{ver}\ and\ var_{ver} < var_{hor})\ or\ ISP_{ver} \\ False, & otherwise \end{cases} \tag{6.1}$$

$$FD_{hor} = \begin{cases} True, & if\ (AIP_{hor}\ and\ var_{hor} < var_{ver})\ or\ ISP_{hor} \\ False, & otherwise \end{cases} \tag{6.2}$$

Figure 6.3 illustrates the flowchart of our scheme, inspired by our previous analysis. For encoding a luminance CTB, the intra-prediction and QTMT partitioning are performed sequentially. If the current CB is partitioned by QT structure, no coding simplification is performed, and the next QT depth is evaluated since the proposed scheme works only in the MTT structure. However, when the current CB is partitioned using the MTT structure, our fast-partitioning decision is analyzed according to the MTT splitting direction. Then, the binary tree partition in horizontal or vertical directions (BTH and BTV) and the ternary tree partition in horizontal or vertical directions (TTH and TTV) are evaluated to be skipped or not. When FD_{hor} is true, the current CB is classified as vertical texture direction, and the horizontal binary and ternary splitting are skipped; otherwise, no simplification is performed, and the next MTT depth is evaluated. FD_{ver} is analogous to FD_{hor} but skipping the vertical splitting.

6.3   Results and Discussion

Our scheme was implemented in VTM 10.0 and evaluated in the all-intra configuration. Its results are displayed in Table 6.1, where the encoding efficiency was measured by the BD-BR metric and encoding time saving (ETS). According to the encoded video sequence, the scheme achieved an average ETS of 28.78%, with an ETS minimum and maximum of 43.23 and 8.12%, respectively. These ETS results were obtained with a small average BD-BR increase of 0.80%, demonstrating that our scheme reduces the encoding time significantly, with a minor impact on encoding efficiency.

The high difference in results obtained in different videos is explained due to their different distribution of ISP and AIP modes. The video sequences with a low selection of ISP and AIP modes, such as ParkRunning3, obtained smaller ETS since our solution takes advantage of these prediction modes to accelerate the encoder. For example, in BasketballDrive, 33.42% of the encoded CBs are predicted with ISP, whereas in ParkRunning3, ISP predicts only 10.21% of the encoded CBs. However, in sequences with shorter time-saving results, the BD-BR loss is smaller.

Table 6.2 displays a summary comparing our scheme with related works [5–9] focusing on accelerating VVC intra-coding partitioning. Our scheme obtained lower ETS than these works, but with better BD-BR results. However, note that our solution only focuses on the MTT structure, whereas three related works focus on the QT and MTT structures. Thus, our scheme can be combined with other techniques focusing on accelerating QT

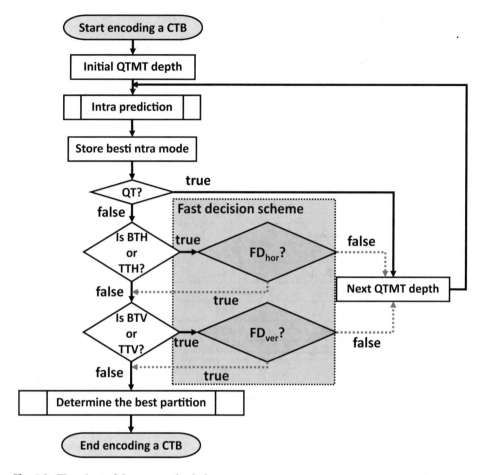

Fig. 6.3 Flowchart of the proposed solution

structure or other encoding modules to provide more impressive ETS results. Neverthe-less, the scheme designed provides competitive results for ETS and BD-BR compared to the related works. It is essential to mention that these related works were evaluated in different VTM versions than ours. Besides, these VTM versions did not include all VVC technologies or several fast decisions that were not available in the very first VTM versions. We evaluated our scheme with VTM 10.0, which encompasses several other fast decisions tools, such as split cost prediction and ternary split restriction, and tools such as MIP and ISP that completely change the behavior of the encoder [10, 11], when compared with the first VTM versions.

Table 6.1 Proposed solution results for CTC evaluation under all-intra configuration

Class	Video sequence	BD-BR (%)	ETS (%)
A1	Tango2	0.30	13.53
	FoodMarket4	0.26	14.57
	Campfire	0.38	12.81
A2	CatRobot	0.72	21.76
	DaylightRoad2	0.92	29.48
	ParkRunning3	0.15	8.12
B	MarketPlace	0.24	14.47
	RitualDance	0.54	25.79
	Cactus	0.78	27.54
	BasketballDrive	0.97	37.46
	BQTerrace	1.33	38.57
C	BasketballDrill	1.37	27.67
	BQMall	1.21	43.23
	PartyScene	0.88	42.31
	RaceHorsesC	0.58	27.71
D	BasketballPass	0.98	37.03
	BQSquare	1.01	40.19
	BlowingBubbles	0.75	35.50
	RaceHorses	0.62	25.98
E	FourPeople	1.11	39.13
	Johnny	1.17	33.52
	KristenAndSara	1.26	36.78
Average		0.80	28.78
Standard deviation (σ)		0.38	10.70

Table 6.2 Comparison of the proposed solution with related works

Work	VTM version	Module	BD-BR (%)	ETS (%)
Our	10.0	MTT	0.80	28.78
Fu et al. [5]	1.0	MTT	1.02	45.00
Yang et al. [6]	2.0	QT + MTT	1.56	52.59
Lei et al. [7]	3.0	MTT	0.84	40.70
Cui et al. [8]	5.0	QT + MTT	1.23	51.01
Zhao et al. [9]	7.0	QT + MTT	0.86	39.39

References

1. Huang, Y., et al. (2021). Block partitioning structure in the VVC standard. *IEEE Transactions on Circuits and Systems for Video Technology, 31*(10), 3818–3833.
2. Chen, J., Ye, Y., & Kim, S. (2020). Algorithm description for versatile video coding and test model 10 (VTM 10). In *JVET 19th Meeting, JVET-S2002, Teleconference.*
3. De-Luxán-Hernández, S., et al. (2019). An intra subpartition coding mode for VVC. In *IEEE International Conference on Image Processing (ICIP)* (pp. 1203–1207).
4. Saldanha, M., et al. (2020). Fast partitioning decision scheme for versatile video coding intra-frame prediction. In *International Symposium on Circuits and Systems (ISCAS)* (pp. 1–5).
5. Fu, T., et al. (2019). Fast CU partitioning algorithm for H.266/VVC intra-frame coding. In *IEEE International Conference on Multimedia and Expo (ICME)* (pp. 55–60).
6. Yang, H., et al. (2020). Low complexity CTU partition structure decision and fast intra mode decision for versatile video coding. *IEEE Transactions on Circuits and Systems for Video Technology (TCSVT), 30*(6), 1668–1682.
7. Lei, M., et al. (2019). Look-ahead prediction based coding unit size pruning for VVC intra coding. In *IEEE International Conference on Image Processing (ICIP)* (pp. 4120–4124).
8. Cui, J., et al. (2020). Gradient-based early termination of CU partition in VVC intra coding. In *Data Compression Conference (DCC)* (pp. 103–112).
9. Zhao, J., et al. (2020). *Adaptive CU split decision based on deep learning and multifeature fusion for H.266/VVC*, (Vol. 2020, pp. 1058–9244). Scientific Programming Hindawi.
10. VTM. (2020). VVC Test Model (VTM). Retrieved October, 2021, from https://vcgit.hhi.fraunhofer.de/jvet/VVCSoftware_VTM/-/releases/VTM-10.0.
11. Pfaff, J., et al. (2021). Intra prediction and mode coding in VVC. *IEEE Transactions on Circuits and Systems for Video Technology (TCSVT), 31*(10), 3834–3847.

Light Gradient Boosting Machine Configurable Fast Block Partitioning for Luminance

7

This chapter enhances the scheme presented in Chap. 6 using a machine learning classifier called Light Gradient Boosting Machine (LGBM), which is used to decide the QTMT block partitioning as a multiple binary classification problem. An offline training was performed, creating one classifier for each split type. The classifiers are used to decide whether to perform the split type, skipping the evaluation of split types that are unlikely to be chosen as the best ones. We selected the LGBM classifier since, during our early stage analysis, it provided higher accuracy than other possibilities. Moreover, LGBM has a high potential for improving performance due to the high flexibility of the hyperparameters configuration. This configurable fast block partitioning scheme was published at IEEE Transactions on Circuits and Systems for Video Technology (TCSVT) [1].

7.1 Background on LGBM Classifiers

Combining several weak machine learning models can be used to improve the overall system performance and provide higher accuracy than individual models [2]. The two main types of ensemble approaches are bagging, which creates individual classifiers for taking decisions based on the majority votes of all classifiers, and boosting, which builds the classifiers iteratively, minimizing the error of the earlier trained classifiers [2].

Microsoft researchers have developed LGBM, a gradient boosting framework that uses tree-based learning algorithms [3]. Figure 7.1 presents the LGBM training approach that builds a decision tree ensemble sequentially to minimize losses and improve the model at each iteration step. A new decision tree model is created in each iteration concerning the error of the entire ensemble learned so far. The learning rate parameter controls the gradient descent approach that minimizes the losses when adding trees.

A solid predictive model is obtained by LGBM combining N tree models ($f_1, f_2, f_3, \ldots, f_n$), and Eq. 7.1 describes the results aggregated from each step.

© The Author(s), under exclusive license to Springer Nature Switzerland AG 2022
M. Saldanha et al., *Versatile Video Coding (VVC)*, Synthesis Lectures on Engineering, Science, and Technology, https://doi.org/10.1007/978-3-031-11640-7_7

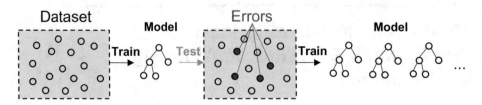

Fig. 7.1 LGBM training approach

$$f(x) = \sum_{n=1}^{N} f_n(x) \qquad (7.1)$$

Unlike other tree-based learning algorithms, LGBM grows trees leaf-wise (vertically) since the prediction loss is reduced by the algorithms that produce level-wise trees (horizontally). Moreover, conventional implementations of Gradient Boosting Machines (GBM) scan all the data instances to estimate the information gain of all possible split points, which requires a significant computational effort during the training process. To overcome this problem, LGBM uses sampling methods for data selections called Gradient-based One-Side Sampling (GOSS) and Exclusive Feature Bundling (EFB). These methods discard some well-trained instances (small training errors) and reduce the dimensionality of the features while maintaining high accuracy [3].

The LGBM technique provides a highly flexible training process to control the learning rate hyperparameters, dataset sampling, and decision tree characteristics, generating a high-efficient model when adequately optimized. LGBM has several advantages compared to other models like:

(i)   Capable of handling large-scale data.
(ii)   Support of parallel and Graphics Processing Unit (GPU) learning.
(iii) Low memory usage.
(iv)   Fast training speed.
(v)   Simple implementation with tree-based algorithm.
(vi)   High accuracy.
(vii) Low inference time.

The characteristics (v), (vi), and (vii) are essential for our work since our goal is to reduce the encoding time without impacting the coding efficiency.

7.2   Methodology

We discovered strong correlations between the coding context and its attributes by applying data mining. We used these correlations for defining machine learning models that determine when to perform a QTMT split type, saving coding time with negligible efficiency reduction. Our solution divides the block partition decision into five binary classification problems instead of creating an LGBM classifier that directly solves the QTMT structure multiclass problem. This approach enables the design of specialized classifiers for each split type, saving expressive encoding time while minimizing the coding efficiency loss. We trained the LGBM classifiers offline for each split type, including QT, BTH, BTV, TTH, and TTV; each classifier decides to skip or not the corresponding split type.

Figure 7.2 illustrates the methodology used to train and implement the LGBM classifiers in the VTM encoder. A set containing eight video sequences was selected for training. The VTM encoder was modified to extract and output several statistical data with relevant information for the CB split decision, generating one dataset for each split type. These datasets contain features that were extracted from the encoded video sequences, encoder attributes, and the split decision for each CB. The datasets were balanced, and the most important features were selected in our preprocessing step.

These selected features were used as input for the model training. The model training includes two steps: hyperparameter optimization and each classifier training. Finally, after the model training, the LGBM classifiers were implemented into the VTM encoder and evaluated by employing test sequences not used in the training phase. Then, the coding efficiency and encoding time saving of the proposed solution can be compared with the unmodified VTM.

The video sequences used during the model training are described in Table 7.1. These sequences were selected with different characteristics and resolutions ranging from 416×240 up to 3840×2160 pixels [4–6]. The video sequences used in the training process

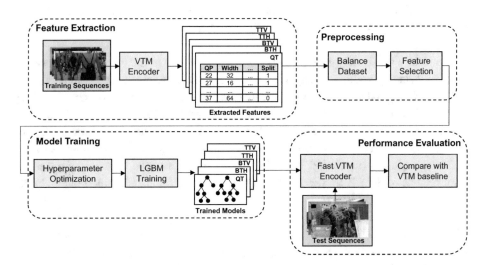

Fig. 7.2 Framework for training CB partitioning decision with LGBM models and evaluating the performance in the VTM encoder

Table 7.1 Video sequences used for training

Training sequence	Resolution	Bit depth	FPS
Traffic flow	3840×2160	10	30
Building hall2	3840×2160	10	50
Kimono1	1920×1080	8	24
Park scene	1920×1080	8	24
Vidyo1	1280×720	8	60
Netflix_drivingPOV	1280×720	8	60
Pedestrian_area	832×480 (downsampled)	8	25
Flowervase	416×240	8	30

encompass a wide range of video characteristics (e.g., 8–10-bit depth and 24–60 frames per second—fps) for rendering several examples of block partitioning decisions in the training process.

During the training, the videos were encoded following the encoder configurations specified in JVET CTC for all-intra configuration, using QP values 22, 27, 32, and 37. However, we limited the encoding to 120 frames to reduce the training complexity.

7.3   Features Analysis and Selection

A large amount of data was extracted from the test video sequences to find features that could lead to effective decisions on the CB split type. These features encompass four information categories: (i) CB Samples, (ii) Local Samples, (iii) Context, and (iv) Coding Information.

The feature set (i) only encompasses the features related to the current luminance CB samples, including width and height of the current CB, area, block ratio, variance (*var*), horizontal (*Gx*), and vertical (*Gy*) gradients based on the Sobel operator, *Gx* divided by *Gy* (*ratioGxGy*), and the sum of *Gx* and *Gy* divided by the block area (*normGradient*).

The feature set (ii) computes the local samples using subparts of the current CB; i.e., the absolute difference of variances on four sub-quarters (*diffVarQT*), maximum variance on four sub-quarters (*maxVarQT*), the absolute difference between variances of upper and lower regions of the CB (*diffVarHor*), and the absolute difference between left and right regions of the CB (*diffVarVer*).

The feature set (iii) computes context information using data of the neighboring CBs. It includes average QT (*neighAvgQT*) and MTT (*neighAvgMTT*) depth levels in neighboring CBs and the number of neighboring CBs with QT (*neighHigherQT*) and MTT (*neighHigherMTT*) depth levels higher than the current CB.

Several results obtained during the evaluation of the current CB size can be used for deciding the split type since no split is assessed before QT, BT, and TT splits. The feature set (iv) includes features related to data obtained in the current CB evaluation— QP, RD cost (*currCost*), distortion (*currDistortion*), current QT (*QTD*), BT (*BTD*), MTT (*MTTD*), and QTMT (*QTMTD*) depth levels, best intra-prediction mode (*currIntraMode*), MRL index (*mrlIdx*), LFNST index (*lfnstIdx*), ISP mode (*ispMode*), and MTS flag (*mts-Flag*). Besides, considering that the split types are evaluated in order, the split decision can consider the previous split information; i.e., BTH RD cost (*costBTH*), BTV RD cost (*costBTV*), *costBTH* divided by *costBTV* (*ratioCostBTHBTV*), and TTH RD cost (*costTTH*). The corresponding RD-cost is unavailable when a previous split evaluation type is skipped; thus, the feature is assigned with the maximum finite double-precision value.

Table 7.2 shows the features selected in the five designed classifiers. These features were determined using the Feature Selector tool [7]. Collinear and low-importance features were removed from the full feature set to reduce the computational effort of the training process.

Figure 7.3 displays the ten features with higher importance for each classifier. They were measured using the split metric, which calculates the number of times the model uses the features. Features related to the RD cost (i.e., *currCost* and *currDistortion*) have high importance for all proposed classifiers. Besides, using the RD cost of previous splits also provides valuable information for the subsequent split evaluations. As expected, features that use a specific direction are highly related to their corresponding split direction (see BTH, TTH, BTV, and TTV in Fig. 7.3).

Figure 7.4 plots the probability density functions of four selected features for QT, BTH, and BTV classifiers. These graphs correlate with the evaluated values and the splitting decision. For example, Fig. 7.4a shows that low values of *currCost* should not perform a QT split.

7.4   Classifiers Training and Performance

The classifier training process requires maximizing the model performance by hyperparameter optimization. LGBM brings several hyperparameters to provide higher accuracy and deal with overfitting and underfitting that need to be properly optimized. Therefore, the hyperparameters of each classifier were optimized using the efficient Optuna framework [8] and applying the Tree-structured Parzen Estimator (TPE) [9] approach.

Table 7.3 displays the main optimized hyperparameters for each classifier. *Learning_rate* corresponds to the speed the error is corrected from each iteration (or tree) to the next. *Feature_fraction* specifies the percentage of features used for each iteration. *Bagging_fraction* specifies the fraction of data (training examples) used for each iteration, whereas *bagging_freq* indicates the frequency k for performing bagging. *Num_leaves*

Table 7.2 Features used for each classifier

Feature	Description	QT	BTH	BTV	TTH	TTV
QP	The current QP value	×	×	×	×	×
currCost	The current RD-cost	×	×	×	×	×
currDistortion	The current distortion	×	×	×	×	×
width	The current block width	×	×	×	×	×
height	The current block height		×	×	×	×
area	The current block area		×	×	×	×
blockRatio	Block width divided by block height		×	×	×	×
QTD	The current QT depth level		×			
BTD	The current BT depth level		×	×	×	×
MTTD	The current MTT depth level		×	×	×	×
QTMTD	The current QTMT depth level		×	×		
currIntraMode	The current intra-prediction mode	×	×	×	×	×
mrlIdx	Reference line index of MRL		×			
ispMode	Identify the ISP mode	×	×	×	×	×
mtsFlag	Identify the use of MTS		×	×	×	×
lfnstIdx	Identify the index of LFNST		×	×		
var	Block variance	×	×	×	×	×
diffVarQT	Absolute difference of variances on four sub-quarters	×	×	×	×	×
maxVarQT	Maximum variance on four sub-quarters	×	×	×	×	×
diffVarHor	Absolute difference among variances of upper and lower CB regions	×	×	×	×	×
diffVarVer	Absolute difference among variances of left and right CB regions	×	×	×	×	×
Gx	Horizontal Sobel gradient	×	×	×	×	×
Gy	Vertical Sobel gradient	×	×	×	×	×
ratioGxGy	Gx divided by Gy	×	×	×	×	×
normGradient	Sum of Gx and Gy divided by block the area	×	×	×	×	×

(continued)

Table 7.2 (continued)

Feature	Description	QT	BTH	BTV	TTH	TTV
neigHavgQT	Average QT depth level in neighboring CBs	×	×	×	×	×
neighHigherQT	Higher QT depth level in neighboring CBs	×	×	×	×	×
neighAvgMTT	Average MTT depth level in neighboring CBs	×	×	×	×	×
neighHigherMTT	Higher MTT depth level in neighboring CBs	×	×	×	×	×
costBTH	RD-cost of BTH split type			×	×	×
costBTV	RD-cost of BTV split type				×	×
ratioCostBTHBTV	RD-cost of BTH divided by RD-cost of BTV				×	×
costTTH	RD-cost of TTH split type					×
Number of features		19	28	29	28	29

denotes the maximum number of leaves in one tree, and *Max_depth* limits the maximum depth for each tree. Finally, *Num_iterations* specifies the number of boosting iterations (or the number of trees).

The obtained classifiers were evaluated using the ten-fold cross-validation. Results of this evaluation are presented in Table 7.4, where accuracy measures the ratio of the correct predictions over the total number of instances evaluated, and the F1-score is the harmonic mean among precision and recall values [10]. Our results demonstrate that the classifiers obtain stable results for both metrics (accuracy and F1-score) and can provide high performance to predict the CB split type.

Our proposed algorithm uses the LGBM classifiers to skip the evaluation of split types with a low probability of being selected as the optimal partitional. Each LGBM classifier indicates a value related to the probability of skipping that evaluation, comparing this value with a threshold. This decision threshold is configurable, where different tradeoff results between encoding time reduction and encoding efficiency can be achieved. By default, the decision threshold used by the LGBM model is 0.5, and the confidence of prediction is given by how close to 0 or 1 is the decision function output. If the output is higher than 0.5, the classifier decides to skip the split type evaluation; otherwise, the classifier remains the split type evaluation.

The performance of our algorithm implemented into VTM is displayed in Fig. 7.5, where seven different thresholds were evaluated: 0.3, 0.4, 0.5, 0.55, 0.6, 0.65, and 0.7. This evaluation allows the analysis of the individual results of each classifier for different operation points to validate their performance in terms of time saving and BD-BR. When smaller threshold values are applied, there is a higher encoding time reduction and

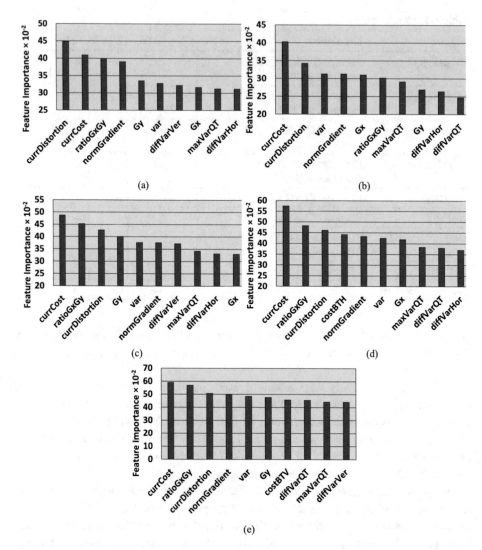

Fig. 7.3 Feature importance ranking of top ten features for **a** QT, **b** BTH, **c** BTV, **d** TTH, and **e** TTV classifiers

a higher increase in BD-BR since more splits are skipped. In contrast, the higher the threshold value, the lower the encoding time reduction and BD-BR impact since more splits are evaluated. Therefore, the threshold values 0.3 and 0.7 provide the highest and the lowest time savings for all classifiers, respectively.

This analysis shows that the classifiers can obtain good results individually; however, integrating all classifiers is not a trivial task, and some adaptations were needed.

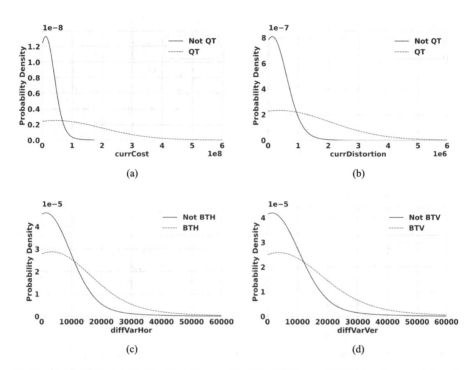

Fig. 7.4 Probability density functions for **a** and **b** QT, **c** BTH, and **d** BTV classifiers regarding four analyzed attributes

Table 7.3 Optimized hyperparameters for each classifier

Hyperparameter	QT	BTH	BTV	TTH	TTV
learning_rate	0.10	0.13	0.12	0.12	0.14
feature_fraction	0.84	0.70	0.81	0.84	0.92
bagging_fraction	0.73	0.71	0.97	0.86	0.98
bagging_freq	3	5	7	7	1
num_leaves	254	250	231	251	253
max_depth	24	60	13	39	34
num_iterations	176	176	256	268	298

Table 7.4 Accuracy and F1-score results for each classifier

Metric	QT (%)	BTH (%)	BTV (%)	TTH (%)	TTV (%)
Accuracy	83.50	74.69	74.77	76.52	76.52
F1-score	83.18	74.64	74.72	76.59	76.58

Fig. 7.5 Encoding time
reduction and coding efficiency
of each classifier for seven
threshold values

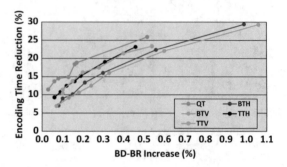

7.5   Classifiers Integration

The flowchart of the complete integrated solution composed of the five LGBM classifiers
is presented in Fig. 7.6. The white and gray colors refer to the native steps of the VTM
encoding flow, and the orange, blue, and pink colors represent the new steps of our solution introduced in the encoder, including feature extraction, classifier evaluation, and split
evaluation decision, respectively.

After evaluating the intra-frame prediction with the not split type, our solution extracts
the features to feed the LGBM classifiers. Then, one LGBM classifier is applied for each
split type that can be evaluated. Given the probability of each classifier compared to a
static threshold, our algorithm decides to evaluate splits with a high probability of being
selected as optimal. Then, the evaluation of a determined split type is skipped if the
probability is higher than the decision threshold; otherwise, the encoding flow remains
without modifications. Our approach follows the same encoding order as the unmodified
VTM; thus, the proposed solution also performs a sequential decision, as shown in the
flowchart. This approach takes advantage of the information of previously evaluated split
types with specialized classifiers for each split type, aiming to increase the accuracy of
the decisions.

Aiming to enhance our results and provide more flexibility, we decided to establish
two thresholds: one for QT, called TH_{QT}, and another one for MTT (including horizontal
and vertical BT/TT partitions), called TH_{MTT}. The result analysis showed that the coding
efficiency is significantly reduced when our solution wrongly decides to skip all MTT
divisions in a given direction. Therefore, we improved our model introducing the possibility of evaluating one split in that given direction. Then, when the model decides that
all splits in a given direction must be skipped and their skip probability is lower than 0.7
(empirically defined), then the split type with the lowest skip probability is evaluated.

By using configurable decision thresholds, our solution can change the operation point
at multiple granularities according to the application requirements, such as levels of CTB,
frame, GOP, or video. All experiments presented in this chapter considered the video level
granularity. The change in the operating point can be performed by changing the TH_{QT}
and TH_{MTT} values, according to the flowchart presented in Fig. 7.6. On the one hand,

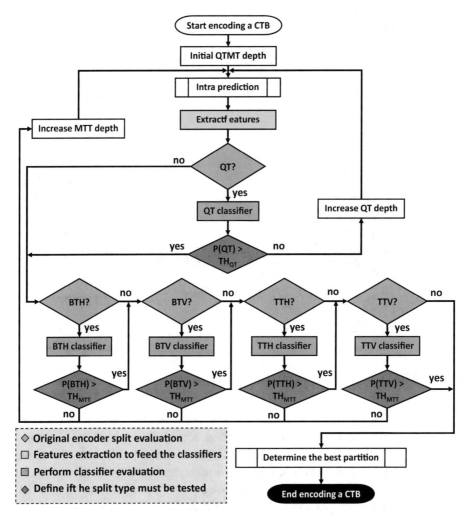

Fig. 7.6 Flowchart of the proposed solution integrated with the QTMT splitting process

when these thresholds are increased, the number of evaluated split types also increases, resulting in a smaller time-saving improvement while maintaining higher coding efficiency. On the other hand, decreasing the threshold values increases the number of split types skipped, resulting in a higher encoding time reduction.

The following section evaluates the proposed solution, including five operation points described in Table 7.5. A dense experimental analysis led to the definition of these operation points, showing promising results to support different application requirements. However, according to the application requirements, our solution is still configurable, allowing the usage of different thresholds than those applied in our evaluation.

Table 7.5 Values of TH_{QT} and TH_{MTT} for the five operation points

Configuration	TH_{QT}	TH_{MTT}
C_1	0.7	0.7
C_2	0.7	0.6
C_3	0.6	0.55
C_4	0.5	0.5
C_5	0.4	0.4

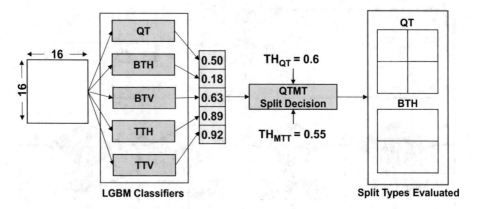

Fig. 7.7 QTMT split decision using the proposed solution for a 16×16 CB

An example of the QTMT split decision using our solution for a 16×16 CB is presented in Fig. 7.7, where the C_3 operation point is used. Each classifier is applied to determine the probability of skipping that split type. Then, BTV, TTH, and TTV decided to skip the evaluations since their probabilities were higher than the thresholds defined in C_3. In this case, only QT and BTH split types are evaluated for the current CB since only these splits have a lower probability than the decision threshold values.

7.6 Results and Discussion

Table 7.6 demonstrates our solution results, including encoding time saving (ETS) and coding efficiency (BD-BR) for five operation points. One can notice that the video sequences evaluated here are the JVET CTC sequences, which are different from the videos used in the training step (see Sect. 7.2). This allowed a robust evaluation of the proposed solution. Our experiments considered the five operation points presented in the previous section and a configuration at the video level, meaning that the operation point was static during the video sequence encoding.

Table 7.6 ETS and coding efficiency results of the proposed solution for five operation points

Class	Video sequence	Configuration level									
		C_1		C_2		C_3		C_4		C_5	
		BD-BR (%)	ETS (%)	BD-BR (%)	ETS (%)	BD-BR (%)	ETS (%)	BD-BR (%)	ETS (%)	BD-BR (%)	ETS (%)
A1	Tango2	0.39	42.53	0.49	48.54	0.71	53.15	1.01	57.19	1.62	61.97
	FoodMarket4	0.35	36.12	0.50	41.28	0.69	46.55	0.96	51.18	1.51	56.82
	Campfire	0.39	29.40	0.59	38.85	0.83	43.76	1.18	51.82	2.02	59.57
A2	CatRobot	0.41	30.39	0.64	38.68	0.90	44.12	1.23	50.81	2.11	57.28
	DaylightRoad2	0.50	38.87	0.84	47.75	1.10	53.67	1.48	59.48	2.45	66.82
	ParkRunning3	0.21	28.70	0.31	35.58	0.45	41.36	0.61	47.60	0.98	52.12
B	MarketPlace	0.23	35.65	0.40	46.98	0.60	54.47	0.84	60.62	1.41	67.65
	RitualDance	0.52	40.04	0.79	46.23	1.09	53.46	1.53	58.88	2.54	65.40
	Cactus	0.45	34.70	0.73	44.42	1.04	49.99	1.47	57.24	2.58	64.76
	BasketballDrive	0.63	44.99	0.94	50.59	1.26	57.12	1.67	60.13	2.61	66.88
	BQTerrace	0.55	32.36	0.84	41.75	1.11	48.35	1.48	54.42	2.36	62.92
C	BasketballDrill	0.51	23.82	0.97	33.72	1.52	40.31	2.40	45.40	4.44	56.30
	BQMall	0.66	39.34	1.03	46.44	1.40	51.05	1.96	56.37	3.27	63.12

(continued)

Table 7.6 (continued)

Class	Video sequence	Configuration level									
		C_1		C_2		C_3		C_4		C_5	
		BD-BR (%)	ETS (%)	BD-BR (%)	ETS (%)	BD-BR (%)	ETS (%)	BD-BR (%)	ETS (%)	BD-BR (%)	ETS (%)
	PartyScene	0.27	35.90	0.52	41.74	0.77	48.33	1.15	53.66	2.11	61.16
	RaceHorsesC	0.34	32.63	0.53	41.36	0.75	46.95	1.07	52.53	1.97	61.01
D	BasketballPass	0.57	38.47	0.86	44.32	1.32	49.27	1.86	53.74	3.19	59.90
	BQSquare	0.21	26.51	0.39	35.61	0.57	40.66	0.85	47.05	1.58	56.36
	BlowingBubbles	0.27	29.87	0.52	38.14	0.82	43.71	1.19	48.98	2.27	57.15
	RaceHorses	0.28	28.52	0.45	37.35	0.72	45.04	1.06	49.90	2.08	56.86
E	FourPeople	0.83	40.33	1.24	47.99	1.71	54.11	2.38	59.09	4.04	66.54
	Johnny	0.97	43.92	1.27	49.98	1.65	55.30	2.13	59.37	3.45	65.65
	KristenAndSara	0.64	41.89	0.97	48.32	1.26	52.96	1.69	56.88	2.86	63.13
Average		0.46	35.22	0.72	42.98	1.01	48.80	1.42	54.20	2.43	61.34
Standard deviation (σ)		0.20	6.00	0.28	5.12	0.36	5.12	0.50	4.65	0.85	4.36

Fig. 7.8 ETS and BD-BR increase for the five operation points of the proposed configurable solution and comparison with the related works

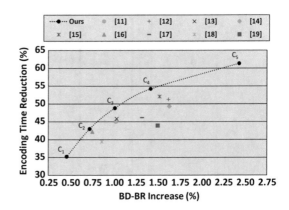

An extensive range of ETS and BD-BR impacts can be noticed in the five operation points (C_1, C_2, ..., C_5), as displayed in Table 7.6. According to the selected operation point, ETS has varied between 35.22% to 61.34%, while BD-BR has increased from 0.46% to 2.43%. Our solution supports various application requirements, with expressive ETS gains and a minor impact on the BD-BR results. Stable results are also obtained in different video sequences since we obtained a low standard deviation for ETS and BD-BR, even considering different video characteristics and resolutions.

Figure 7.8 summarizes the comparisons with related works, where the relation between ETS and BD-BR of each work and operation points are plotted. We included in this comparison nine related works [11–19].

C_1–C_5 identify the five operation points of our configurable solution in Fig. 7.8. The dotted line is obtained through interpolation of the operation points, where those values are extended for identifying other possibilities for operation points (i.e., with different thresholds).

Our solution surpasses all related works since it achieved a higher ETS for a similar BD-BR impact. Figure 7.8 also clarifies the high flexibility level of our configurable method compared to the related works since our work can explore different relations between ETS and BD-BR according to the application requirements.

A deeper comparison with these related works is presented in Table 7.7, where the average BD-BR and ETS results for each video class are presented. Only C_3 and C_4 operational points are displayed in this table since they are the most comparable with the related works. We used the related works [11, 12, 15, 17] in this comparison since they provide detailed results and used almost the same experimental setup considered in our work, making this comparison fairer.

One can notice operation point C_3 can reach the smallest BD-BR results among all related works, and a higher ETS than the solution of Fu et al. [11] and Li et al. [17]. Besides, C_3 achieved the smallest standard deviation results for BD-BR among all related works, and a smaller standard deviation for ETS than Yang et al. [15] and Chen et al.

Table 7.7 Comparison with related works

| Class | [11] | | [12] | | [15] | | [17] | | This work | | | |
| | | | | | | | | | C_3 | | C_4 | |
	BD-BR (%)	ETS (%)	BD-BR (%)	ETS (%)	BD-BR (%)	ETS (%)	BD-BR (%)	ETS (%)	BD-BR (%)	ETS (%)	BD-BR (%)	ETS (%)
A1	1.31	51.00	1.17	42.71	0.85	51.39	1.60	43.90	0.74	47.82	1.05	53.39
A2	1.19	47.67	1.60	48.36	0.77	53.08	1.49	45.48	0.82	46.38	1.11	52.63
B	0.92	47.60	1.56	51.96	2.09	58.91	1.15	49.09	1.02	52.68	1.40	58.26
C	0.98	42.25	1.63	53.79	1.48	49.39	1.09	45.18	1.11	46.66	1.65	51.99
D	0.62	40.75	1.30	53.86	1.19	44.16	1.07	43.03	0.86	44.67	1.24	49.92
E	1.31	40.00	2.55	54.96	2.85	58.60	1.81	49.50	1.54	54.12	2.07	58.44
Avg	1.02	45.00	1.62	51.23	1.52	52.01	1.32	46.13	1.01	48.80	1.42	54.20
σ	0.41	4.94	0.58	6.27	0.85	6.47	0.45	3.87	0.36	5.12	0.50	4.65

[12]. Operation point C_4 achieved the highest ETS among all related works and a lower BD-BR than the works of Chen et al. [12] and Yang et al. [15]. This configuration also reached lower standard deviation results for ETS than the works [11, 12, 15] and a lower BD-BR standard deviation than works [12, 15].

The reached results outperform these related works in terms of combined rate-distortion and time saving.

References

1. Saldanha, M., et al. (2021). Configurable fast block partitioning for VVC intra coding using light gradient boosting machine. *IEEE Transactions on Circuits and Systems for Video Technology (TCSVT).*
2. Natekin, A., & Knoll, A. (2013). Gradient boosting machines, a tutorial. *Frontiers in Neuro-robotics, 7*, 21.
3. Ke, G., et al. (2017). LightGBM: A highly efficient gradient boosting decision tree. *Advances in Neural Information Processing Systems, 30*, 3146–3154.
4. Sharman, K., & Suehring, K. (2017). Common test conditions. JCT-VC 26th meeting, JCTVC-Z1100.
5. Daede, T., Norkin, A., & Brailovskiy, I. (2019). Video Codec testing and quality measurement. Draft-ietf-netvc-testing-08 (work in progress), p. 23.
6. Xiph. Xiph.org video test Media [derf's collection]. Xiph.org. Retrieved October, 2021, from https://media.xiph.org/video/derf/.
7. Koehrsen, W. (2021) GitHub-WillKoehrsen/feature-selector. Feature selector, 2018. Retrieved October, 2021, from https://github.com/WillKoehrsen/feature-selector.
8. Akiba, T., et al. (2019) Optuna: A next-generation hyperparameter optimization framework. In *Proceedings of the 25th ACM SIGKDD International Conference on Knowledge Discovery & Data Mining*, pp. 2623–2631.
9. Bergstra, J., et al. (2011). Algorithms for hyper-parameter optimization. *Neural Information Processing Systems Foundation (NIPS), 24*, 2546–2554.
10. Hossin, M., & Sulaiman, N. (2015). A review on evaluation metrics for data classification evaluations. *International Journal of Data Mining & Knowledge Management Process (IJDKP), 5*(2), 1.
11. Fu, T., et al. (2019). Fast CU partitioning algorithm for H.266/VVC intra-frame coding. In *IEEE International Conference on Multimedia and Expo (ICME)*, pp. 55–60.
12. Chen, F., et al. (2020). A fast CU size decision algorithm for VVC intra prediction based on support vector machine. *Multimedia Tools and Applications (MTAP), 79*, 27923–27939.
13. Lei, M., et al. (2019). Look-ahead prediction based coding unit size pruning for VVC intra coding. In *IEEE International Conference on Image Processing (ICIP)*, pp. 4120–4124.
14. Fan, Y., et al. (2020). A fast QTMT partition decision strategy for VVC intra prediction. *IEEE Access, 8*, 107900–107911.
15. Yang, H., et al. (2020). Low complexity CTU partition structure decision and fast intra mode decision for versatile video coding. *IEEE Transactions on Circuits and Systems for Video Technology (TCSVT), 30*(6), 1668–1682.
16. Tissier, A., et al. (2020). CNN oriented complexity reduction of VVC intra encoder. In *IEEE International Conference on Image Processing (ICIP)*, pp. 3139–3143.

17. Li, T., et al. (2021). DeepQTMT: A deep learning approach for fast QTMT-based CU partition of intra-mode VVC. *IEEE Transactions on Image Processing (TIP), 30,* 5377–5390.
18. Zhao, J., et al. (2020). Adaptive CU split decision based on deep learning and multifeature fusion for H.266/VVC. *Scientific Programming Hindawi, 2020,* 1058–9244.
19. Li, Y., et al. (2021). Early intra CU size decision for versatile video coding based on a tunable decision model. *IEEE Transactions on Broadcasting (TBC), 67*(3), 710–720.

Learning-Based Fast Decision for Intra-frame Prediction Mode Selection for Luminance

This chapter describes the design of a learning-based fast decision scheme to discard some intra-frame prediction modes from the RD-list before the RDO process. This solution was published at IEEE Visual Communications and Image Processing (VCIP) [1].

The proposed scheme is composed of three solutions:

(i)   A fast Planar/DC decision based on a decision tree classifier.
(ii) A fast MIP decision based on a decision tree classifier.
(iii) A fast ISP decision based on the block variance.

Chapter 5 demonstrated the mode selection distribution between the conventional intra-frame prediction approach (i.e., AIP + MRL [2,3]) and the new intra-coding tools (MIP [4] and ISP [5]) can be explored to develop an efficient encoding time-saving solution. On average, AIP plus MRL is assigned 68% of the time, followed by MIP and ISP, respectively, with 24% and 8%. Moreover, Fig. 8.1 shows that planar and DC, which are included in the AIP, are the most used prediction modes in the conventional intra-frame prediction approach (AIP + MRL), representing 43% of the occurrences. Considering only angular modes, the most used one (mode 50) occurs less than 5% of the time.

The conventional intra-frame prediction approach provides the best encoding possibility for most cases, frequently using Planar or DC modes. Consequently, an efficient scheme to decide when avoiding angular, MIP, and/or ISP modes evaluations in the RDO process can save high encoding time with negligible coding efficiency reduction. The proposed scheme also applies data mining to discover strong correlations between the encoding context and its attributes, but for defining the decision tree classifiers that decide when avoiding some intra-frame prediction modes evaluations. We further implemented VTM functions and collected statistical information from the encoding process to create the datasets balanced according to the number of instances for each frame, block

M. Saldanha et al., *Versatile Video Coding (VVC)*, Synthesis Lectures on Engineering, Science, and Technology, https://doi.org/10.1007/978-3-031-11640-7_8

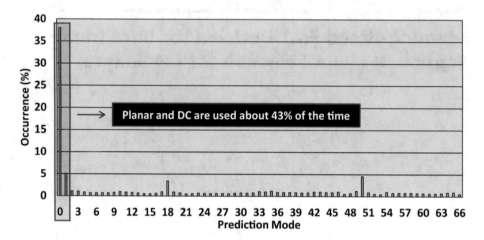

Fig. 8.1 Mode selection distribution for conventional intra-frame prediction approach

size, QP value, and output class. The decision trees were trained offline using the REP-Tree algorithm in the *Waikato Environment for Knowledge Analysis* (WEKA) [6]. The training process used the same methodology (training sequences, QP values, and VTM configurations) detailed in Chap. 7.

8.1 Fast Planar/DC Decision Based on Decision Tree Classifier

The first solution is a decision tree classifier to choose when planar and DC are likely chosen as the best prediction modes. In this case, our solution removes the angular modes from the RD-list, avoiding their evaluations in the RDO process. We collected many features from the encoding process to define this decision tree and selected the best ones. Table 8.1 presents the features used in the decision tree classifier with the corresponding descriptions and Information Gain (IG) [7].

Figures 8.2 a and b demonstrate the correlation of the *smoothIsFirst* and *numAng-Modes* features with the smooth mode decision, respectively. *Smooth* and *Not Smooth* indicate when Planar or DC and angular modes are selected as the best ones, respectively. Figure 8.2 displays that these features are highly correlated with the smooth mode decision. For instance, when *smoothIsFirst* is 1, a smooth mode is selected at about 91%. Also, the smaller the value of *numAngModes,* the higher the chance of choosing a smooth mode.

Planar/DC decision tree was designed with an 8-depth and 85-size tree, providing 85.37% accuracy (F1-score 85.4%) when evaluated with ten-fold cross-validation.

Table 8.1 Features used in the Planar/DC decision tree classifier

Feature	Description	IG
QP	The current QP value	0.094
area	The current block area	0.040
dcInRdList	Notify if DC is on RD-list	0.147
posPlanar	The position of the Planar in the RD-list	0.309
posDC	The position of the DC in the RD-list	0.270
dcIsMPM	Notify if DC is an MPM	0.130
smoothIsFirst	Notify if Planar or DC is on the first RD-list position	0.214
mipIsFirst	Notify if MIP mode is on the first RD-list position	0.031
numAngModes	Number of angular modes in the RD-list	0.350
numMipModes	Number of MIP modes in the RD-list	0.174
numModesList	Total number of modes in the RD-list	0.130
dcSATDCost	SATD-cost of DC	0.281
planarSATDCost	SATD-cost of Planar	0.192
firstSATDNoSmooth	SATD-cost of the first angular or MIP in the RD-list	0.005
ratioSATDSmooth	SATD-cost of Planar divided by SATD-cost of DC	0.094
ratioSATD	Minimum SATD-cost between Planar and DC divided by firstSATDNoSmooth	0.201

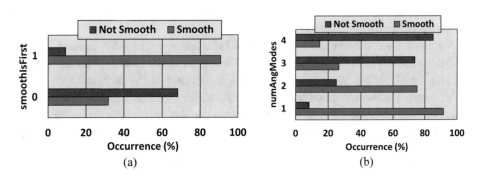

Fig. 8.2 Correlation of *smoothIsFirst* and *numAngModes* features with the smooth mode decision

8.2 Fast MIP Decision based on Decision Tree Classifier

Since MIP modes are used 24% of the time (see Chap. 5), our second solution decides when MIP modes can be removed from the RD-list to avoid their evaluations in the RDO process. This solution also uses a decision tree classifier to verify when a MIP mode is

Table 8.2 Features used in the MIP decision tree classifier

Feature	Description	IG
QP	The current QP value	0.093
Width	The current block width	0.081
Height	The current block height	0.080
Area	The current block area	0.038
blockRatio	Block width divided by block height	0.055
QTMTD	The current QTMT depth level	0.109
posPlanar	The position of the Planar in the RD-list	0.138
posFirstMip	The position of the first MIP in the RD-list	0.295
numMipModes	Number of MIP modes in the RD-list	0.135
numModesList	Total number of modes in the RD-list	0.129
planarSATDCost	SATD-cost of Planar	0.005
firstSATDMip	SATD-cost of the first MIP in the RD-list	0.005
ratioConvMip	SATD-cost of the first conventional mode divided by the first MIP in the RD-list	0.230
numNeighMip	Number of neighboring blocks encoded with MIP	0.057

unlikely to be chosen as the best mode. Once again, we collected many new features from the encoding process and selected those with higher IG according to the MIP decision. Table 8.2 presents the features in the MIP decision tree classifier with the corresponding descriptions and IG.

Figures 8.3 a and b correlate the *numMipModes* and *ratioConvMIP* features with the MIP decision, respectively. *MIP* and *Not MIP* denote when the block is encoded with and without a MIP mode, respectively. Figure 8.3 illustrates that the probability of a block to be encoded with MIP raises with the *numMipModes* increase. Besides, low values of *ratioConvMip* or values close to one tend to use a non-MIP mode, whereas values higher than one tend to select a MIP mode. Both features show a high correlation with the MIP decision.

We designed the MIP decision tree with an 8-depth and 97-size tree, obtaining accuracy results of 78.83% (F1-score 78.8%) when evaluated using ten-fold cross-validation.

8.3   Fast ISP Decision Based on the Block Variance

Our third solution is a fast ISP decision based on the block variance to avoid ISP evaluations in the RDO process. The solution is based on the fact that ISP modes are frequently used to predict more complex texture regions where the conventional prediction approach cannot provide high coding efficiency. Nevertheless, the traditional prediction approach

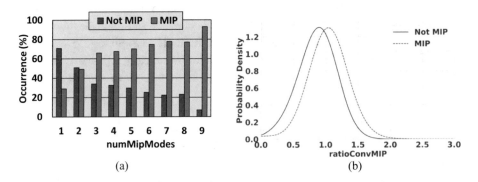

Fig. 8.3 Correlation of *numMipModes* and *ratioConvMIP* features with the MIP decision

can reduce rate distortion for simpler texture regions. Thus, the fast ISP decision analyzes the block texture complexity to decide when to remove ISP modes from the RD-list, skipping their evaluations in the RDO process.

Figure 8.4 depicts the probability density functions for ISP mode selection according to the block variance, considering Vidyo1 (Fig. 8.4a) and Kimono1 (Fig. 8.4b) video sequences encoded with QP 32; *ISP* and *Not ISP* denote when a block is encoded with and without ISP, respectively.

Figure 8.4 shows that the block variance provides a high correlation to define when using ISP modes, and a threshold decision can be defined to remove ISP modes from the RD-list. This figure also shows that a static threshold definition may sometimes deal with inaccurate decisions since it tends to change according to the video content and quantization scenario. Consequently, the proposed solution uses block variance to measure the block texture complexity and decide when avoiding ISP modes evaluations in the RDO, but the threshold value is computed online during the encoding of the first frame of

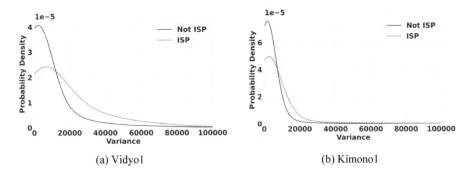

Fig. 8.4 Probability density functions for ISP mode selection according to the block variance of two video sequences

the video sequence. For this purpose, the solution stores the variance values of all blocks of the first frame that did not select ISP as the best mode and computes the average variance value, used in the subsequent frames as a threshold to avoid the ISP evaluations in the RDO. Additionally, the threshold value can be periodically adjusted according to the application requirements regarding the number of frames, video content change, or target encoding time reduction.

8.4   Learning-Based Fast Decision Design

Figure 8.5 shows the flowchart of the learning-based fast decision scheme designed to reduce the encoding effort of the intra-frame prediction mode selection for luminance CBs. Our decisions are carried out before the RDO process and after the definition of the RD-list with RMD. The decision of the Planar/DC classifier is computed to decide if angular modes should be removed or not from RD-list. In contrast, the MIP classifier decision is computed to determine if MIP modes should be removed or not from RD-list. The RDO process is simplified, reducing the encoding effort if any classifier decides to release modes from the RD-list.

After that, for the first frame of the video sequence, the ISP modes are evaluated by RDO, and our scheme performs the threshold calculation. In the remaining frames, the fast ISP decision is applied by comparing the variance of the block with the threshold value to decide whether the ISP assessment can be skipped.

8.5   Results and Discussion

This section presents the results of the proposed scheme for VVC intra-coding. It is important to highlight that VTM implements some native speedup heuristics for the intra-frame prediction process, such as a fast RMD search of the 65 angular modes, the MRL evaluation only for MPMs, and fast decisions for ISP based on the previous prediction mode evaluations, as detailed in [8]. Our experiments enable all these speedup techniques for a fairer comparison with the current VTM encoder implementation.

Table 8.3 depicts the ETS and BD-BR results of Planar/DC and MIP DTs, fast ISP decision, and the overall proposed scheme. Planar/DC and MIP DTs provide 10.47% of ETS with a 0.29% BD-BR increase. The fast ISP decision reaches 8.32% of ETS with a 0.31% BD-BR increase. The overall proposed scheme achieves 18.32% of ETS with a negligible impact on BD-BR of 0.60%. Moreover, this scheme presents a small standard deviation result, showing stable results for different video characteristics.

This scheme was the first one published presenting a complexity reduction solution for VVC intra-coding considering all the standardized intra-frame prediction tools; hence, it was difficult to perform a fair comparison with related works. The works of Yang et al. [9]

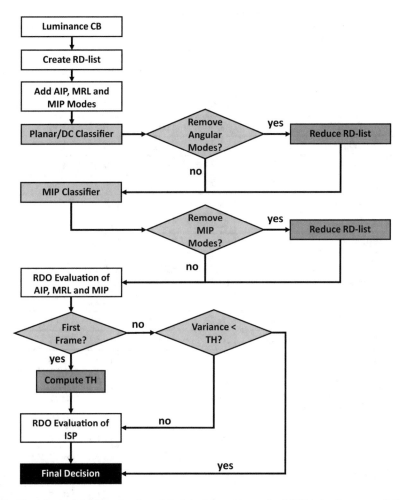

Fig. 8.5 Flowchart of the learning-based fast decision scheme for VVC intra-frame prediction

and Chen et al. [10] used an old version of VTM (2.0) and reached 25.51% of ETS with a 0.54% BD-BR increase and 30.59% of ETS with a 0.86% BD-BR increase, respectively. Since our scheme targeted VTM with all standardized tools (e.g., MIP and ISP), having a more complex intra-coding flow, one can conclude that the proposed scheme reaches competitive results in terms of encoding time reduction versus the coding efficiency losses.

Table 8.3 Proposed scheme results for CTC under all-intra configuration

Class	Video sequence	Planar/DC and MIP decision trees		Fast ISP		Overall	
		BD-BR (%)	ETS (%)	BD-BR (%)	ETS (%)	BD-BR (%)	ETS (%)
A1	Tango2	0.50	11.90	0.08	7.92	0.56	18.89
	FoodMarket4	0.29	8.60	0.07	6.58	0.36	15.29
	Campfire	0.26	11.82	0.03	9.26	0.29	19.63
A2	CatRobot	0.33	9.94	0.24	7.01	0.58	17.62
	DaylightRoad2	0.39	12.80	0.24	9.23	0.64	19.71
	ParkRunning3	0.13	8.15	0.07	5.24	0.18	12.63
B	MarketPlace	0.11	12.47	0.08	10.67	0.22	20.84
	RitualDance	0.19	9.92	0.24	6.84	0.44	19.79
	Cactus	0.26	12.05	0.29	5.94	0.55	19.36
	BasketballDrive	0.42	10.20	0.43	7.95	0.84	19.36
	BQTerrace	0.36	10.29	0.37	10.22	0.72	20.07
C	BasketballDrill	0.41	6.25	0.69	7.99	1.11	16.05
	BQMall	0.33	10.31	0.66	9.43	0.94	19.40
	PartyScene	0.19	11.26	0.34	11.18	0.54	20.28
	RaceHorsesC	0.15	11.47	0.26	11.90	0.41	18.17
D	BasketballPass	0.36	9.98	0.41	7.46	0.72	16.08
	BQSquare	0.32	10.52	0.39	8.25	0.69	19.24
	BlowingBubbles	0.23	9.96	0.34	7.09	0.57	16.55
	RaceHorses	0.15	11.02	0.28	11.38	0.43	19.86
E	FourPeople	0.28	11.48	0.46	6.65	0.79	18.21
	Johnny	0.45	9.61	0.48	6.92	0.85	17.30
	KristenAndSara	0.38	10.32	0.43	7.83	0.83	18.75
Average		0.29	10.47	0.31	8.32	0.60	18.32
Standard deviation (σ)		0.11	1.50	0.18	1.85	0.24	1.98

References

1. Saldanha, M., et al. (2021). Learning-based complexity reduction scheme for VVC intra-frame prediction. In *Proceedings of Visual Communications and Image Processing (VCIP)*, pp. 1–5.
2. Chen, J., Ye, Y., & Kim, S. (2020). Algorithm description for versatile video coding and test model 10 (VTM 10). In *JVET 19th Meeting, JVET-S2002, Teleconference*.

3. Chang, Y., et al. (2019). Multiple reference line coding for most probable modes in intra prediction. *In Proceedings of Data Compression Conference (DCC)*, pp. 559–559.

4. Schafer, M., et al. (2019). An affine-linear intra prediction with complexity constraints. In *Proceedings of IEEE International Conference on Image Processing (ICIP)*, pp. 1089–1093.

5. De-Luxán-Hernández, S., et al. (2019). An intra subpartition coding mode for VVC. In *Proceedings IEEE International Conference on Image Processing (ICIP)*, pp. 1203–1207.

6. Hall, M., et al. (2009). The weka data mining software: An update. *ACM SIGKDD Explorations Newsletter, 11*(1), 10–18.

7. Cover, T., & Thomas, J. (1991). *Elements of information theory.* Wiley.

8. De-Luxán-Hernández, S., et al. (2020). Design of the intra subpartition mode in VVC and its optimized encoder search in VTM. Applications of digital image processing XLIII, 11510, 115100Y.

9. Yang, H., et al. (2020). Low complexity CTU partition structure decision and fast intra mode decision for versatile video coding. *IEEE Transactions on Circuits and Systems for Video Technology (TCSVT), 30*(6), 1668–1682.

10. Chen, Y., et al. (2020). A novel fast intra mode decision for versatile video coding. *Journal of Visual Communication and Image Representation (JVCIR)*, 71, 102849.

Fast Intra-frame Prediction Transform for Luminance Using Decision Trees

9

This chapter presents the development of a fast intra-frame prediction transform scheme to decide when avoiding MTS [1] and LFNST [2] evaluations in the costly RDO process. This fast transform scheme was published at IEEE International Symposium on Circuits and Systems (ISCAS) [3], including two solutions:

(i) A fast MTS decision based on a decision tree classifier.
(ii) A fast LFNST decision based on a decision tree classifier.

Chapter 5 demonstrated that the MTS transform matrices are selected only 35% of the time, indicating that DCT-II and TSM are used in most cases. The distribution presents a slight variation according to the block size but with a low standard deviation. Besides, LFNST is used in 49% of the blocks, indicating that more than 50% of the cases are encoded without LFNST. In this case, different block sizes present a higher variation in the results. The experimental results show that the lowest and the highest percentage of LFNST use occur with 64×64 (33%) and 16×16 (66%) CBs, respectively. These results show that, in most cases, the blocks are encoded without MTS and/or LFNST coding tools, allowing to avoid MTS and LFSNT evaluations in several cases in the costly RDO process. Consequently, the proposed fast transform scheme can provide interesting time-saving results with negligible impact on coding efficiency.

To design this scheme, we used the same methodology applied in Chap. 8, but here the decision tree classifiers determine when avoiding MTS and/or LFNST coding tool evaluations.

© The Author(s), under exclusive license to Springer Nature Switzerland AG 2022   99
M. Saldanha et al., *Versatile Video Coding (VVC)*, Synthesis Lectures on Engineering,
Science, and Technology, https://doi.org/10.1007/978-3-031-11640-7_9

9.1   Fast MTS Decision Based on Decision Tree Classifier

The coding information obtained before MTS evaluation can be used to decide when the MTS evaluation can be avoided since MTS is evaluated later than DCT-II and TSM in the VTM implementation [4, 5]. Therefore, the first solution presented in this chapter is a decision tree classifier that uses encoding context and current coding information to identify when the MTS can be removed from the RDO process. We collected many features from the encoding process to define this decision tree and selected the best ones. Table 9.1 shows the features used in the decision tree classifier with the corresponding descriptions and IGs.

Figure 9.1a and b demonstrate the correlations of the MTS decision with *num-NeighMTS* and *ispMode*, respectively. In these figures, the name *Others* indicates blocks encoded with DCT-II or TSM, and *MTS* denotes blocks encoded with the MTS coding tool. On the one hand, when *numNeighMTS* is zero, about 61% of the blocks are encoded without MTS. On the other hand, the higher the *numNeighMTS*, the higher the probability of encoding the block with MTS. When *ispMode* is one or two, more than 62% of the blocks are encoded without MTS. In contrast, 51% of the blocks are encoded with MTS when this value is zero. These figures demonstrated that both features correlate with the MTS decision.

The MTS decision tree was designed with an 8-depth and 125-size tree that provides 73.87% of accuracy (F1-score 73.8%) when evaluated with ten-fold cross-validation.

Table 9.1 Features used in the MTS decision tree classifier

Feature	Description	IG
QP	The current QP value	0.016
area	The current block area	0.086
width	The current block width	0.116
BTD	The current binary tree depth	0.111
mipFlag	Notify if MIP was selected	0.009
ispMode	Identify the ISP mode	0.017
currCost	RD-cost	0.054
currDistortion	Total distortion	0.073
currFracBits	Number of encoded bits	0.090
currIntraMode	Intra prediction mode	0.068
noIspCost	RD-cost of the best non-ISP mode	0.117
numNonZeroCoeffs	Number of non-zero coefficients	0.026
absSumCoeffs	Absolute sum of the coefficients	0.094
numNeighMTS	Number of neighboring blocks encoded with MTS	0.030

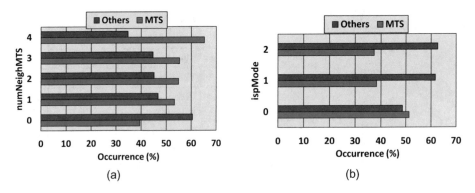

Fig. 9.1 Correlation of *numNeighMTS* and *ispMode* features with the MTS decision

9.2   Fast LFNST Decision Based on Decision Tree Classifier

The second solution is a decision tree classifier to identify when skipping LFNST evaluations since more than 50% of the CBs encoded with DCT-II do not use LFNST. To define this decision tree, we also collected many features from the encoding process when the DCT-II is evaluated without LFNST. We trained the decision tree classifier to decide when the LFNST evaluations can be removed from the RDO process based on the collected data. Table 9.2 shows the features used in this decision tree classifier with the corresponding descriptions and IGs [6].

Figure 9.2a and b exemplify the correlation of *MTTD* and *ispMode* features with the LFNST decision, respectively. When *MTTD* is zero, about 69% of the blocks are encoded without LFNST, whereas for higher values of *MTTD*, the percentage of blocks encoded with LFNST is about 52% (*MTTD* = 4). When *ispMode* is one or two, more than 67% of the blocks are encoded without LFNST, whereas for *ispMode* equals zero, 50% of the blocks are encoded with LFNST, demonstrating a correlation with the LFNST decision.

The LFNST decision tree was designed with an 8-depth and 123-size tree, obtaining accuracy results of 76.75% (F1-score 76.7%) when evaluated using a ten-fold cross-validation.

9.3   Fast Transform Decision Design

Figure 9.3 presents the flowchart of the fast transform decision scheme using decision tree classifiers to reduce the encoding time of the VVC transform coding flow. In the VTM implementation, the encoder evaluates DCT-II (without LFNST) and TSM after creating the RD-list with the best intra-prediction mode candidates. Subsequently, based on the result of the transform with the lowest RD-cost, our scheme computes the features for the

Table 9.2 Features used in the LFNST decision tree classifier

Feature	Description	IG
Width	The current block width	0.019
Height	The current block height	0.170
Area	The current block area	0.055
Block ratio	Block width divided by block height	0.021
MTTD	The current multi-type tree depth	0.176
mipFlag	Notify if MIP was selected	0.106
ispMode	Identify the ISP mode	0.015
currCost	RD-cost	0.086
currDistortion	Total distortion	0.136
currFracBits	Number of encoded bits	0.087
currIntraMode	Intra prediction mode	0.013
numNonZeroCoeffs	Number of non-zero coefficients	0.189
absSumCoeffs	Absolute sum of the coefficients	0.046

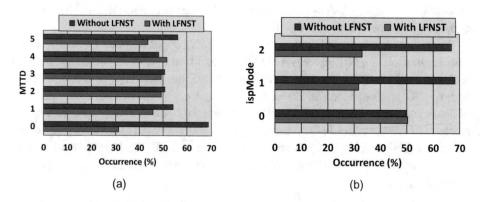

Fig. 9.2 Correlation of *MTTD* and *ispMode* features with the LFNST decision

MTS and LFNST decision tree classifiers and the choice of the MTS and LFNST decision trees. The RDO encoding effort is reduced when the MTS and/or LFNST decision tree chooses to skip the MTS and/or LFNST evaluation; otherwise, the encoding flow remains without modifications.

Fig. 9.3 Flowchart of the
transform decision scheme

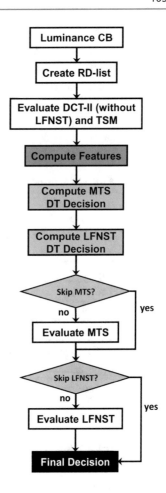

9.4   **Results and Discussion**

This section presents the results of the proposed fast transform decision scheme for VVC
intra-coding of luminance CBs. It is important to highlight that VTM also implements
some native speedups heuristics for the transform coding process, including deciding to
skip the MTS and LFNST evaluations based on the ISP results, and to terminate the
MTS evaluation based on the results of a certain MTS mode, as described in [4]. Our
experiments enable all these speedup techniques, allowing a fairer comparison with the
used VTM encoder implementation.

Table 9.3 shows the ETS and BD-BR results of the MTS and LFNST decision tree
and the overall proposed scheme. The MTS decision tree provides 5% of ETS with a
0.21% BD-BR increase. The LFNST decision tree attains 6.40% of ETS at the cost of
a 0.23% BD-BR increase. The overall proposed scheme reaches 10.99% of ETS with

a negligible BD-BR increase of 0.43%. The proposed scheme presents a low standard deviation, indicating stable results for different video content and resolutions. The BD-BR and ETS results range from 0.15% to 0.70% and from 7.81% to 13.30%, respectively.

This scheme was the first encoding time reduction solution considering all the novel transform coding tools for VVC intra-frame prediction, hampering a fair comparison with related works at that time. The work of Fu et al. [7] focused on an old VTM software implementation (version 3.0) and attained 23% of ETS with a 0.16% BD-BR increase.

Table 9.3 Proposed scheme results for CTC under all-intra configuration

Class	Video sequence	MTS decision tree		LFNST decision tree		Overall	
		BD-BR (%)	ETS (%)	BD-BR (%)	ETS (%)	BD-BR (%)	ETS (%)
A1	Tango2	0.22	6.04	0.47	6.29	0.70	11.83
	FoodMarket4	0.27	2.72	0.41	5.26	0.68	8.91
	Campfire	0.17	3.25	0.16	4.21	0.32	8.24
A2	CatRobot	0.24	5.65	0.18	6.75	0.40	9.94
	DaylightRoad2	0.20	6.21	0.19	3.88	0.40	11.92
	ParkRunning3	0.21	2.76	0.12	6.67	0.30	7.81
B	MarketPlace	0.24	5.29	0.14	5.54	0.38	13.17
	RitualDance	0.27	5.37	0.32	7.62	0.60	7.96
	Cactus	0.23	3.89	0.19	3.90	0.40	11.52
	BasketballDrive	0.27	5.45	0.27	6.85	0.49	11.40
	BQTerrace	0.11	4.45	0.16	5.61	0.25	11.75
C	BasketballDrill	0.25	6.03	0.47	4.64	0.64	9.98
	BQMall	0.25	7.13	0.17	8.45	0.41	10.44
	PartyScene	0.10	5.93	0.08	9.58	0.20	12.70
	RaceHorsesC	0.17	4.81	0.18	7.10	0.32	13.14
D	BasketballPass	0.27	5.50	0.20	6.72	0.45	9.51
	BQSquare	0.09	4.39	0.07	6.68	0.15	13.30
	BlowingBubbles	0.14	5.20	0.12	7.81	0.26	12.20
	RaceHorses	0.17	4.91	0.17	8.64	0.34	11.74
E	FourPeople	0.33	4.29	0.36	7.51	0.66	11.23
	Johnny	0.28	4.67	0.33	4.60	0.58	10.49
	KristenAndSara	0.21	6.13	0.35	6.53	0.59	12.55
Average		0.21	5.00	0.23	6.40	0.43	10.99
Standard deviation (σ)		0.06	1.14	0.12	1.57	0.16	1.70

Nevertheless, this solution reduces the number of intra-prediction modes in the RD-list and does not consider the LFNST coding tool in the encoding flow. Our scheme can provide high encoding time reduction with a minimal impact on the coding efficiency since it targets VTM with all standardized coding tools (e.g., LFNST, MIP, and ISP), presenting a more complex evaluation process. Additionally, our scheme can be combined with other solutions to reduce the number of intra-prediction modes in the RD-list or fast CB decisions to provide even more impressive time-saving results.

References

1. Zhao, X., et al. (2016). Enhanced multiple transform for video coding. In *Proceedings of Data Compression Conference (DCC)* (pp. 73–82).
2. Koo, M., et al. (2019). Low frequency non-separable transform (LFNST). In *Proceedings of IEEE Picture Coding Symposium (PCS)* (pp. 1–5).
3. Saldanha, M., et al. (2022). Fast transform decision scheme for VVC intra-frame prediction using decision trees. In *Proceedings of International Symposium on Circuits and Systems (ISCAS)*.
4. Zhao, X., et al. (2021). Transform coding in the VVC standard. *IEEE Transactions on Circuits and Systems for Video Technology (TCSVT), 31*(10), 3878–3890.
5. Chen, J., Ye, Y., & Kim, S. (2020). Algorithm description for versatile video coding and test model 10 (VTM 10). In *JVET 19th Meeting, JVET-S2002, Teleconference*.
6. Cover, T., & Thomas, J. (1991). *Elements of information theory*. Wiley.
7. Fu, T., et al. (2019). Two-stage fast multiple transform selection algorithm for VVC intra coding. In *Proceedings of IEEE International Conference on Multimedia and Expo (ICME)* (pp. 61–66).

Heuristic-Based Fast Block Partitioning Scheme for Chrominance

10

This chapter describes a heuristic-based fast block partitioning scheme for chrominance intra-coding to predict the QTMT depth level and split direction. This fast block partitioning scheme for chrominance coding was published in the SPIE Journal of Electronic Imaging (JEI) [1].

This scheme is composed of two heuristics:

(i) Chrominance CB splitting early termination based on luminance QTMT.
(ii) Fast chrominance split decision based on the variance of sub-blocks.

Even though the highest encoding effort required is related to the luminance blocks, the encoding effort of chrominance blocks cannot be neglected, mainly for real-time applications. Therefore, we developed a fast block partitioning scheme for chrominance intra-coding, exploring the correlation between chrominance and luminance coding structures and the statistical information in the chrominance block samples.

Figure 10.1a, b and c show the block size distributions for luminance (Y), chrominance blue (Cb), and chrominance red (Cr), respectively. These block size distributions were extracted from the first frame of the BasketballPass test sequence encoded with all-intra configuration and QP 37. The same block size distribution is obtained for Cb and Cr since VVC jointly encodes these components [2, 3]. In most cases, chrominance is encoded with larger block sizes than luminance since chrominance present more homogeneous regions. Nevertheless, the chrominance QTMT structure is evaluated considering all splitting possibilities to find the best one, requiring a high encoding effort [3].

In I-slices, luminance and chrominance QTMT structures are independently generated for each CTU [2], but each chrominance CTB encoding is carried out after encoding the associated luminance CTB. Since these components represent the same scene, it is possible to explore some correlations between their QTMT structures. Moreover, the statistical

M. Saldanha et al., *Versatile Video Coding (VVC)*, Synthesis Lectures on Engineering, Science, and Technology, https://doi.org/10.1007/978-3-031-11640-7_10

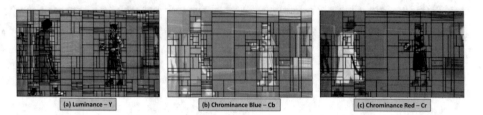

Fig. 10.1 Block size distribution for **a** luminance (Y); **b** chrominance blue (Cb); and **c** chrominance red (Cr)

information of the chrominance CB samples can also be used to identify the texture behavior and predict the chrominance split type.

10.1 Chrominance CB Splitting Early Termination Based on Luminance QTMT

Our first heuristic proposes to explore the correlation of the luminance and chrominance QTMT structures. For a given chrominance CB being evaluated in the QTMT depth d, it is possible to verify the split type used in the associated luminance CB for the same QTMT depth d. Based on this fact, the premise of this heuristic is that if the used split type for the luminance CB in this QTMT depth is not split, there is a high chance that the best chrominance CB has been found at a depth lower or equal to d. Thus, the chrominance block partitioning evaluation can be early terminated.

Figure 10.2 presents the success rate of this approach for five video sequences with resolutions ranging from 416×240 to 3840×2160 [4–6] regarding the four CTC QP values.

The success rate, computed as defined in (10.1), refers to the number of cases for which the proposed predictor had success divided by the total number of cases.

$$Success\ Rate = \frac{Predictor\ have\ success}{Total\ cases}(\%) \tag{10.1}$$

The total number of cases is the number of chrominance blocks evaluated. The predictor has success when the chrominance block is encoded with a QTMT depth lower or equal to the QTMT depth of the associated luminance block; on average, this predictor achieves a success rate higher than 70% for all QPs. The success rate increases as QP increases, obtaining up to 81% for QP 37. On average, this approach provides a 75% of success rate, demonstrating that the luminance QTMT structure is a good predictor for the chrominance QTMT coding.

Fig. 10.2 Success rate of the best chrominance block size found in the QTMT depth lower or equal to luminance QTMT

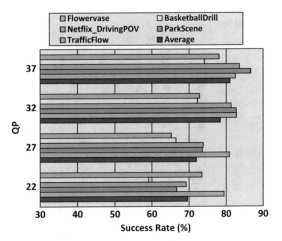

In addition to this evaluation, Fig. 10.3 shows the probability density functions for "split" and "not split" curves according to the RD-cost divided by the block area. This analysis considers the same set of video sequences as the previous evaluation. We decided to compute the RD-cost divided by the block area since larger blocks tend to have larger RD-costs. In this case, the block area denotes the number of samples inside the block, and it is calculated by multiplying the block width by the block height. The chrominance RD-cost was collected for both curves when the collocated luminance CB was defined as not split. The "split" denotes that the chrominance CB was divided when the collocated luminance CB was defined as not split. The "not split" indicates that the chrominance CB was not divided when the collocated luminance CB also was defined as not split.

This analysis exhibits a high probability of success of the QTMT luminance predictor for low RD-cost while having almost no chance of success for higher values. Therefore, this approach can be used to develop an efficient early termination decision based on a threshold criterion, improving the success rate and reducing the coding efficiency loss.

10.2 Fast Chrominance Split Decision Based on Variance of Sub-blocks

The analysis presented in Fu et al. [7] shows that BT and TT split direction is highly linked to the block texture direction because the best block partitioning is obtained with a split type that results in a more homogeneous region or a region with similar behavior. It raises the efficiency of the prediction tools and reduces the residual error, incurring a smaller RD-cost. Consequently, the second heuristic explores the variance of sub-blocks in the current encoding chrominance CB to decide between horizontal and vertical BT or TT splits.

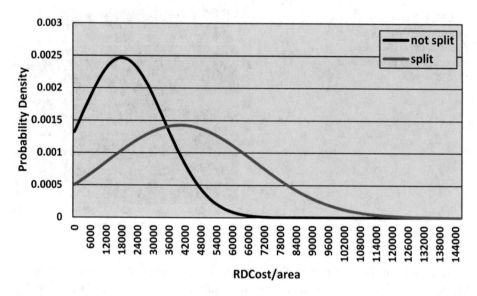

Fig. 10.3 Probability density function of splitting or not the chrominance CB using RD-cost based on the luminance QTMT

Figure 10.4 illustrates the calculation of horizontal and vertical variances for an 8×8 chrominance CB. Figure 10.4a displays the horizontal variance computation—the heuristic considers that the block is horizontally subdivided into two equal-sized sub-blocks (in a similar way as BTH splitting), resulting in two 8×4 sub-blocks. The var_{upper} and var_{lower} values are obtained by calculating the variances of the highlighted regions in red and green, respectively. Thus, var_{hor} denotes the sum of the upper (var_{upper}) and lower (var_{lower}) partitions. Analogously, var_{ver} considers that current CB is vertically subdivided (in a similar way as BTV splitting) and var_{left} and var_{right} are summed to compute var_{ver}.

Considering that the lowest sum of variances indicates that the split type supplies more homogeneous regions, the vertical splitting evaluation could be avoided in Fig. 10.4 since horizontal splitting is the most promising to attain a better rate-distortion result.

VTM encoder minimizes the RD-cost of chrominance blocks considering both chrominance components Cb and Cr to define the best encoding tools possibilities [8]. Hence, the proposed predictor in this heuristic should consider the features of both chrominance components to provide a more accurate decision. For this purpose, this heuristic computes the var_{hor} and var_{ver} separately for Cb and Cr and only skips a given BT/TT direction if both chrominance components agree to skip that direction. For example, the vertical splitting is skipped only if var_{hor} (Cb) is lower than var_{ver} (Cb) and var_{hor} (Cr) is lower than var_{ver} (Cr). For simplifying, we use only var_{hor} and var_{ver} to represent the condition of Cb and Cr.

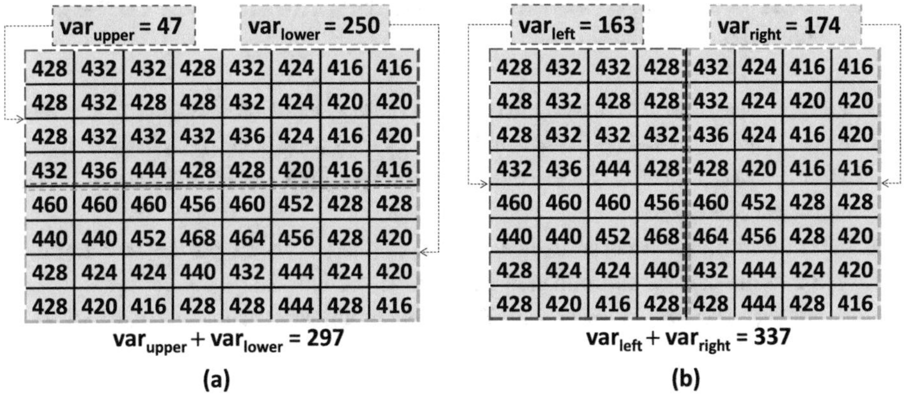

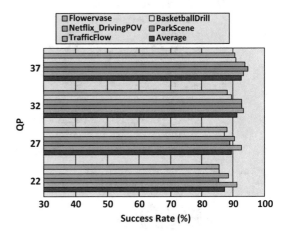

Fig. 10.4 Image of **a** horizontal and **b** vertical variance calculations for an 8×8 chrominance block

Figure 10.5 shows the success rate of skipping the vertical splits when var_{hor} is lower than var_{ver} and skipping the horizontal splits when var_{ver} is lower than var_{hor}. Following Eq. 10.1, the predictor succeeds if it skips the vertical direction and the best split type selected is neither BTV nor TTV or if it skips the horizontal direction and the best split type selected is neither BTH nor TTH. The total number of cases is the total number of chrominance blocks that evaluated at least one horizontal and one vertical split.

This predictor reaches more than 85% of success for all cases evaluated. On average, this predictor provides a success rate of 90%, indicating that this heuristic can provide excellent time-saving results without significant coding efficiency loss.

Fig. 10.5 Success rate of skipping horizontal or vertical splitting of BT and TT partitions using the variance of sub-blocks

10.3   Fast Block Partitioning Scheme for Chrominance Coding Design

Figure 10.6 presents the flowchart of the heuristic-based fast block partitioning scheme for chrominance coding, including the Chrominance CB Splitting Early Termination based on Luminance QTMT (CSETL) and the Fast Chrominance Split Decision based on Variance of Sub-blocks (FCSDV).

CSETL terminates the chrominance QTMT evaluation based on the luminance CB split type and the current chrominance RD-cost. If the luminance CB is "not split" and the chrominance RD-cost is lower than TH_1, the chrominance CB is not partitioned and

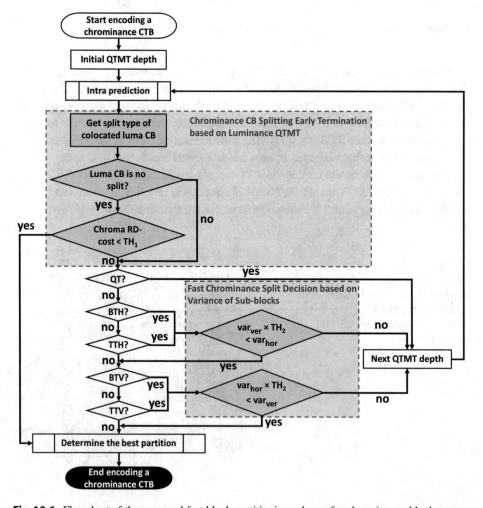

Fig. 10.6 Flowchart of the proposed fast block partitioning scheme for chrominance blocks

the QTMT evaluation is terminated; otherwise, the process continues to evaluate other splitting possibilities.

FCSDV decides the direction of BT and TT splitting based on the variance of sub-blocks. On the one hand, when var_{hor} is lower than var_{ver}, the vertical split types are skipped; on the other hand, when var_{ver} is lower than var_{hor}, the horizontal split types are skipped. Nevertheless, we also decided to introduce another threshold criterion (TH_2) in this heuristic because very close variance values have no obvious texture direction, making it difficult the decision on the best split direction. Then, TH_2 is a value that indicates how many percent a variance value of a given direction should be less than the variance value of the other direction to skip the evaluation.

To define these threshold values, we performed a threshold evaluation considering five scenarios for each heuristic of the proposed scheme. This evaluation assesses the same video sequences presented in Sect. 10.1. The results are presented in terms of chrominance encoding time saving (C-ETS) and BD-BR.

It is important to mention that BD-BR can be calculated individually for each component (Y, Cb, and Cr) or combined for the three components. For all cases, the overall bit rate is used to compute BD-BR (i.e., the bits considering the three components). Nevertheless, different values of Peak Signal to Noise Ratio (PSNR) are used to compute the luminance (Y), chrominance blue (Cb), and chrominance red (Cr) BD-BR results, according to the corresponding component. YCbCr-BDBR also uses the overall bit rate, but it computes PSNR considering the three components through a weighted PSNR average [9], providing the coding efficiency result considering all components together. Since our scheme focuses only on chrominance coding in addition to the chrominance BD-BR results (Cb-BDBR and Cr-BDBR), we showed the complete YCbCr-BDBR to evaluate the quality impact in all video sequence components and the total bit rate, as recommended in [9].

Figure 10.7 shows the C-ETS and BD-BR results collected for different threshold values of TH_1 in the CSETL heuristic. As previously discussed, TH_1 is defined based on the RD-cost divided by the block area. The evaluation scenarios for each threshold TH_1 were defined as follows.

$$TH_1 = \mu + k \times \sigma \qquad (10.2)$$

where k is empirically defined (ranging from 0 to 4), and σ and μ denote the standard deviation and average values, respectively. Figure 10.7 displays that CSETL allows different operation points to improve the coding efficiency or increase time saving. The TH1b detached with a solid red circle reaches a good tradeoff between C-ETS and BD-BR; however, TH1d highlighted with a dotted blue circle indicates the highest C-ETS. Thus, we selected these two interesting points as case studies in the experimental results presented in Sect. 10.4.

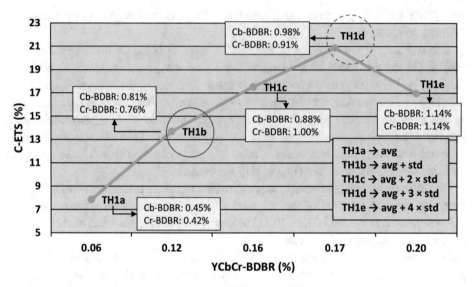

Fig. 10.7 Chrominance encoding time saving and BD-BR impact for CSETL solution according to some threshold values

Figure 10.8 presents the BD-BR and C-ETS results for TH_2 in the FCSDV heuristic. Section 10.3 demonstrated a high success rate for this heuristic; then, we decided to analyze the impact of using some threshold values and also without using a threshold (denoted as "noTH" in Fig. 10.8). These threshold values define how many percent one variance must be less than the other for skipping the BT and TT direction. For example, $TH_2 = 105\%$ defines that a given direction variance must be less than 5% of the other for skipping a direction evaluation.

We also selected two case studies for this heuristic. On the one hand, TH2a, highlighted with a solid red circle in Fig. 10.8, provides a good tradeoff between BD-BR and C-ETS, since it can avoid the cases where variance values are very close. On the other hand, "noTH" obtains the highest C-ETS. Combined with TH1d (CSETL heuristic), the latter can provide the highest ETS of the proposed scheme.

Thus, we defined two case studies regarding two threshold combinations to evaluate in the next section. Case 1 denotes the combination of TH1b (CSETL) and TH2a (FCSDV), which achieves the best tradeoff between time-saving and coding efficiency. Case 2 refers to the combination of TH1d (CSETL) and "noTH" (FCSDV), which provides the highest time saving of this scheme.

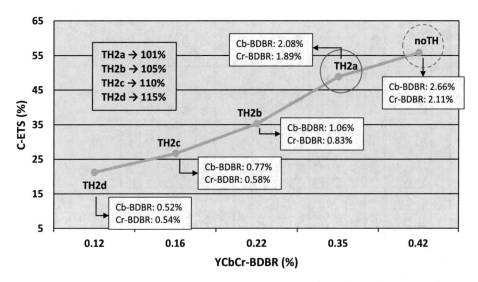

Fig. 10.8 Chrominance encoding time saving and BD-BR impact for FCSDV solution according to some threshold values

10.4 Results and Discussion

This section presents the results of the heuristic-based fast block partitioning scheme for chrominance intra-coding encompassing CSETL and FCSDV heuristics. Table 10.1 depicts the proposed scheme results for Case 1 and Case 2. T-ETS denotes the total time savings (including luminance and chrominance coding), and C-ETS only indicates the time savings for chrominance coding. "Average (without *)" refers to the average of all video sequences, except BasketballDrill, used in the data collection step.

Case 1 is the threshold configuration that reaches a better tradeoff between C-ETS and coding efficiency, saving about 60.03% and 8.18% of C-ETS and T-ETS, respectively, with a minimal impact of 0.66% on YCbCr-BDBR (average without *). The proposed scheme presents a negligible variation in the average results when BasketballDrill is considered, indicating that this scheme can provide excellent results regardless of the video sequence characteristics. Campfire and ParkRunning3 test sequences present the highest (72.97%) and lowest (42.90%) results considering C-ETS, respectively. Regarding YCbCr-BDBR, Campfire and BQSquare produce the highest (1.79%) and the lowest (0.20%) YCbCr-BDBR increase, respectively. These results showed that the proposed scheme reduces more than 60% of the chrominance encoding time with negligible impact on the coding efficiency.

Note that the highest C-ETS does not always result in the highest T-ETS since the chrominance and luminance encoding effort changes according to the video sequence and

Table 10.1 Experimental results obtained with proposed fast block partitioning scheme for chrominance coding under all-intra configuration

Class	Video sequence	Case 1					Case 2				
		YCbCr BDBR (%)	Cb BDBR (%)	Cr BDBR (%)	C-ETS (%)	T-ETS (%)	YCbCr BDBR (%)	Cb BDBR (%)	Cr BDBR (%)	C-ETS (%)	T-ETS (%)
A1	Tango2	1.07	7.91	8.40	64.29	4.87	1.11	8.26	8.69	65.75	5.72
	FoodMarket4	0.76	3.37	3.71	62.84	6.85	0.82	3.69	3.99	66.23	6.93
	Campfire	1.79	5.96	8.35	72.97	14.73	2.33	8.54	9.30	81.18	17.15
A2	CatRobot	1.54	7.70	7.32	68.73	11.76	1.93	9.51	9.12	74.77	13.16
	DaylightRoad2	0.39	4.98	3.74	58.52	4.77	0.46	6.07	4.32	64.08	6.56
	ParkRunning3	0.45	0.90	0.85	42.90	14.04	0.90	1.94	1.75	66.73	21.59
B	MarketPlace	0.55	3.90	2.95	60.49	6.97	0.72	5.15	3.56	68.86	8.95
	RitualDance	0.62	3.53	3.91	59.37	5.79	0.70	3.98	4.50	64.29	5.88
	Cactus	0.55	2.97	3.67	58.03	5.59	0.70	3.76	4.67	65.02	5.07
	BasketballDrive	0.64	3.17	3.70	58.42	4.11	0.75	3.78	4.35	63.71	6.23
	BQTerrace	0.21	3.12	2.92	53.95	6.24	0.28	3.72	3.54	61.61	7.09
C	BasketballDrill *	1.06	3.91	4.18	59.76	9.63	1.29	4.70	5.01	66.07	10.83
	BQMall	0.62	3.34	3.46	63.08	9.64	0.71	3.68	4.11	67.11	9.78
	PartyScene	0.39	2.35	2.53	57.50	10.02	0.50	2.95	3.26	63.94	11.59
	RaceHorsesC	0.52	1.61	2.43	64.20	10.78	0.67	1.81	2.97	70.57	12.28

(continued)

Table 10.1 (continued)

Class	Video sequence	Case 1					Case 2				
		YCbCr BDBR (%)	Cb BDBR (%)	Cr BDBR (%)	C-ETS (%)	T-ETS (%)	YCbCr BDBR (%)	Cb BDBR (%)	Cr BDBR (%)	C-ETS (%)	T-ETS (%)
D	BasketballPass	0.88	3.47	3.08	59.75	8.05	1.08	4.02	3.76	65.91	9.60
	BQSquare	0.20	1.67	2.42	53.18	5.16	0.22	1.76	2.69	58.80	5.77
	BlowingBubbles	0.42	1.80	2.33	57.61	9.74	0.49	2.48	2.61	62.86	9.18
	RaceHorses	0.71	2.18	3.04	62.47	10.28	0.85	2.56	3.82	69.80	11.82
E	FourPeople	0.36	1.74	1.91	57.23	6.22	0.41	1.90	2.33	60.96	6.25
	Johnny	0.72	3.73	2.26	62.51	6.51	0.76	3.92	2.74	65.99	8.07
	KristenAndSara	0.52	2.46	2.21	62.65	9.64	0.58	2.70	2.59	65.82	8.79
Average (without *)		0.66	3.42	3.58	60.03	8.18	0.81	4.10	4.22	66.38	9.40
σ (without *)		0.39	1.83	1.96	5.86	2.99	0.50	2.19	2.12	4.75	4.04

quantization scenario. For example, MarketPlace obtained 60.49% of C-ETS and 6.97% of T-ETS, whereas ParkRunning3 achieved 49.90% of C-ETS and 14.04% of T-ETS.

Case 2 is the threshold configuration that provides a higher C-ETS than others. This configuration obtains C-ETS and T-ETS of 66.38% and 9.40%, respectively, with a 0.81% BD-BR increase. In this case, a higher time saving was achieved at the cost of a YCbCr-BDBR rise if compared to Case 1. Considering only the chrominance components, Case 1 impacts 3.42% and 3.58% on Cb-BDBR and Cr-BDBR, whereas Case 2 increases the Cb-BDBR and Cr-BDBR by 4.10% and 4.22%, respectively.

When published, this was the first solution to reduce the encoding effort of chrominance block partitioning in the VVC encoder, demonstrating the obtained results of chrominance time-saving and coding efficiency.

References

1. Saldanha, M., et al. (2021). Fast block partitioning scheme for chrominance intra prediction of versatile video coding standard. *Journal of Electronic Imaging (JEI), 30*(5), 053009.
2. Pfaff, J., et al. (2021). Intra prediction and mode coding in VVC. *IEEE Transactions on Circuits and Systems for Video Technology (TCSVT), 31*(10), 3834–3847.
3. Huang, Y., et al. (2021). Block partitioning structure in the VVC standard. *IEEE Transactions on Circuits and Systems for Video Technology, 31*(10), 3818–3833.
4. Sharman, K., & Suehring, K. (2017). Common test conditions. In *JCT-VC 26th Meeting, JCTVC-Z1100.*
5. Daede, T., Norkin, A., & Brailovskiy, I. (2019). Video codec testing and quality measurement. draft-ietf-netvc-testing-08 (work in progress), p. 23.
6. Xiph. Xiph.org Video Test Media [derf's collection]. Xiph.org. https://media.xiph.org/video/derf/. Accessed on: Oct 2021.
7. Fu, T., et al. (2019). Fast CU partitioning algorithm for H.266/VVC intra-frame coding. In *Proceedings of IEEE International Conference on Multimedia and Expo (ICME)*, pp. 55–60.
8. Chen, J., Ye, Y., & Kim, S. (2020). Algorithm description for versatile video coding and test model 10 (VTM 10). In *JVET 19th Meeting, JVET-S2002, Teleconference.*
9. Itu, T. (2020). Hstp-Vid-Wpom–working practices using objective metrics for evaluation of video coding efficiency experiments. Technical Paper.

Conclusions and Open Research Possibilities 11

This book proposed several algorithms using heuristic and machine learning approaches at different granularity levels to reduce the encoding effort in the VVC intra-frame prediction. This book was motivated by the high computational requirements of the VVC encoder in improving coding efficiency by adopting several novel complex coding tools. The VVC standardization proposed and adopted several new coding tools to handle high video resolution efficiently, increasing the encoding effort. Hence, VVC introduces new challenges regarding efficient real-time video processing, requiring efficient techniques to reduce its computational cost.

This book includes time-saving algorithms with negligible impact on the coding efficiency for the most time-consuming VVC intra-frame prediction modules, contributing to providing VVC real-time encoding. The main novelties of this book are the detailed discussion of the novel VVC coding tools, the deep analysis of intra-frame prediction steps, and the solutions based on heuristic and machine learning techniques to reduce the encoding effort of intra-frame prediction.

Chapters 2 and 3 present the main VVC novelties. Chapter 2 presents an overview of this new standard, considering some basic video coding concepts, the main steps of a hybrid encoder like the VVC, the frame organization and block partitioning, the main novelties in each encoder step and the VVC common test conditions. Chapter 3 presents details about the VVC intra-frame prediction, which is the focus of this book. In this case, the coding tools are detailed, highlighting the novel VVC intra-frame prediction tools, and the encoding flow for luminance and chrominance samples are also presented. Finally, the transform processing for VVC intra-frame prediction is also detailed.

Chapter 4 presents a summary of the most relevant works in the literature targeting the VCC intra-frame prediction, highlighting the main ideas explored in those works and the reached results.

© The Author(s), under exclusive license to Springer Nature Switzerland AG 2022   119
M. Saldanha et al., *Versatile Video Coding (VVC)*, Synthesis Lectures on Engineering, Science, and Technology, https://doi.org/10.1007/978-3-031-11640-7_11

Chapter 5 presents an extensive performance analysis of intra-frame prediction, encompassing an encoding time and usage distribution analysis. This analysis enables us to identify the most time-consuming modules of VVC intra-frame prediction, which is decisive in developing our encoding time reduction solutions. Moreover, the results presented in this chapter proved that VVC reaches outstanding coding efficiency gain over HEVC.

Chapters 6 and 7 present a heuristic-based solution and a configurable solution based on machine learning for encoding effort reduction of the block partitioning structure of luminance samples. The heuristic-based solution decides when to avoid horizontal or vertical splitting evaluations using the block samples variance and the current prediction mode. This solution reaches 29% of encoding time saving with a 0.8% BD-BR increase. The configurable solution based on machine learning defines an LGBM classifier for each split type, which decides if a given splitting evaluation can be skipped. This solution is configurable and provides different operation points to reduce from 35.22 to 61.34% of the encoding time, affecting only 0.46–2.43% of the BD-BR.

Chapters 8 and 9 englobe the intra-frame prediction and transform selection using decision tree classifiers. The intra-mode selection comprises two solutions using decision trees and one using an online heuristic decision based on statistical analysis. The decision trees decide when removing angular and MIP modes from the RDO evaluation. In contrast, the heuristic determines when skipping ISP evaluations in the RDO process based on the block samples variance. This scheme obtains an 18.32% encoding time reduction with a 0.60% BD-BR increase. The transform selection also encompasses two solutions using decision trees to decide when to avoid MTS and LFNST evaluations. This scheme saves 11% encoding time with a 0.43% BD-BR increase.

Chapter 10 presents the heuristic-based fast block partitioning scheme for chrominance blocks. This scheme explores the correlation between chrominance and luminance blocks and computes the chrominance block sample variances to avoid some split evaluations in the chrominance coding process. This scheme provides more than 60% of chrominance encoding time reduction, increasing only about 0.7% in BD-BR. This was the first solution in the literature targeting chrominance block partitioning.

In summary, this book presented the design of several algorithms and improvements in the VVC encoder by applying heuristic and machine learning techniques. The experimental evaluations showed that significant encoding time reduction was attained by applying these approaches with a minor impact on the coding efficiency. Nevertheless, several points can still be explored to achieve a higher time saving in the VVC encoder. For instance, higher reductions can be achieved by investigating novel features that are more meaningful to the transform or mode selection and the design of new solutions for the mode selection of chrominance blocks.

Furthermore, the literature has not covered several research possibilities yet in the VVC encoder. Few works in the literature focus on the VVC inter-frame prediction, a module with high computational complexity in modern video coders. For instance, the block partitioning structure in inter-frame prediction has not been explored enough. VVC also introduces novel coding tools for inter-frame prediction, including geometric partitioning mode and affine motion compensation, which introduce several new challenges and have not yet been explored sufficiently. Considering hardware design, only a few works designed solutions for VVC coding tools and most of these works are focused on transform coding. However, there is still room to design fast and low-power systems for intra- and inter-frame prediction tools. Besides, there are few works exploring tiles for parallelism in the VVC encoder.

Finally, many video content is still encoded with previous standards such as H.264 and HEVC, demanding the transcoding to VVC format to enable gradual migration to the most recent video coding standard. Consequently, efficient encoding time reduction solutions are also required for transcoding H.264 or HEVC contents to the VVC format.

Index

A
Affine motion compensation, 15, 121

C
Chrominance coding, 107, 112, 113, 115, 116, 120
Configurable solution, 85, 120

D
Decision tree, 4, 35, 71, 72, 89–92, 96, 99–102, 104, 105, 120

E
Encoding effort reduction, 120

H
Hardware design, 121
Heuristics, 3, 4, 24, 35, 36, 63, 94, 103, 107–115, 119, 120

I
Inter-frame prediction, 3, 8–10, 12–15, 17, 18, 120, 121
Intra-frame prediction, 3, 4, 9–12, 14–16, 18, 23–32, 32, 35, 36, 38–40, 43–46, 48–51, 59, 60, 63, 65, 80, 89, 90, 94, 95, 99, 104, 119, 120
Intra sub-partition, 14, 24, 28, 29, 31, 46, 48, 50, 51, 53, 58, 59, 63, 65–68, 75, 76, 89, 92–96, 100, 102, 103, 105, 120

L
Light Gradient Boosting Machine (LGBM), 4, 71–73, 75, 77, 80, 120
Low-frequency non-separable transform, 16, 24, 30, 32, 46, 48, 51, 53, 55, 56, 58–60, 75, 76, 99, 101–105, 120

M
Machine learning, 4, 27, 35, 71, 72, 119, 120
Matrix-based intra prediction, 14, 24, 27–30, 45, 48, 50, 59, 68, 89, 91–96, 100, 102, 105, 120
Multiple reference line, 14, 24, 26
Multiple transform selection, 16, 24, 30
Multi-type tree, 11, 12, 23, 37, 39, 40, 46, 56, 58, 60, 63–67, 69, 74–77, 80, 102

P
Performance analysis, 3, 43, 120

Q
Quadtree with nested Multi-Type Tree (QTMT), 11, 12, 26, 45, 58, 60, 66, 71–73, 75, 76, 81, 82, 92, 107–110, 112, 113

V
Versatile Video Coding (VVC), 1–4, 7, 8, 10–18, 23–32, 35–39, 43–46, 48, 51–60, 67, 68, 94, 95, 101, 103, 104, 107, 118–121
Video coding, 1–3, 7, 8, 119, 121

© The Editor(s) (if applicable) and The Author(s), under exclusive license   123
to Springer Nature Switzerland AG 2022
M. Saldanha et al., *Versatile Video Coding (VVC)*, Synthesis Lectures on Engineering,
Science, and Technology, https://doi.org/10.1007/978-3-031-11640-7

Printed in the United States
by Baker & Taylor Publisher Services